DATA SCIENCE FOUNDATIONS

Geometry and Topology of Complex Hierarchic Systems and Big Data Analytics

Chapman & Hall/CRC
Computer Science and Data Analysis Series

The interface between the computer and statistical sciences is increasing, as each discipline seeks to harness the power and resources of the other. This series aims to foster the integration between the computer sciences and statistical, numerical, and probabilistic methods by publishing a broad range of reference works, textbooks, and handbooks.

SERIES EDITORS

David Blei, Princeton University
David Madigan, Rutgers University
Marina Meila, University of Washington
Fionn Murtagh, Royal Holloway, University of London

Proposals for the series should be sent directly to one of the series editors above, or submitted to:

Chapman & Hall/CRC
Taylor and Francis Group
3 Park Square, Milton Park
Abingdon, OX14 4RN, UK

Published Titles

Semisupervised Learning for Computational Linguistics
Steven Abney

Visualization and Verbalization of Data
Jörg Blasius and Michael Greenacre

Design and Modeling for Computer Experiments
Kai-Tai Fang, Runze Li, and Agus Sudjianto

Microarray Image Analysis: An Algorithmic Approach
Karl Fraser, Zidong Wang, and Xiaohui Liu

R Programming for Bioinformatics
Robert Gentleman

Exploratory Multivariate Analysis by Example Using R
François Husson, Sébastien Lê, and Jérôme Pagès

Bayesian Artificial Intelligence, Second Edition
Kevin B. Korb and Ann E. Nicholson

Published Titles cont.

Computational Statistics Handbook with MATLAB®, Third Edition
Wendy L. Martinez and Angel R. Martinez

Exploratory Data Analysis with MATLAB®, Third Edition
Wendy L. Martinez, Angel R. Martinez, and Jeffrey L. Solka

Statistics in MATLAB®: A Primer
Wendy L. Martinez and MoonJung Cho

Clustering for Data Mining: A Data Recovery Approach, Second Edition
Boris Mirkin

Introduction to Machine Learning and Bioinformatics
Sushmita Mitra, Sujay Datta, Theodore Perkins, and George Michailidis

Introduction to Data Technologies
Paul Murrell

R Graphics
Paul Murrell

Correspondence Analysis and Data Coding with Java and R
Fionn Murtagh

Data Science Foundations: Geometry and Topology of Complex Hierarchic
Systems and Big Data Analytics
Fionn Murtagh

Pattern Recognition Algorithms for Data Mining
Sankar K. Pal and Pabitra Mitra

Statistical Computing with R
Maria L. Rizzo

Statistical Learning and Data Science
*Mireille Gettler Summa, Léon Bottou, Bernard Goldfarb, Fionn Murtagh,
Catherine Pardoux, and Myriam Touati*

Music Data Analysis: Foundations and Applications
Claus Weihs, Dietmar Jannach, Igor Vatolkin, and Günter Rudolph

Foundations of Statistical Algorithms: With References to R Packages
Claus Weihs, Olaf Mersmann, and Uwe Ligges

Chapman & Hall/CRC
Computer Science and Data Analysis Series

DATA SCIENCE FOUNDATIONS

Geometry and Topology of Complex Hierarchic Systems and Big Data Analytics

Fionn Murtagh

CRC Press
Taylor & Francis Group
Boca Raton London New York

CRC Press is an imprint of the
Taylor & Francis Group, an **informa** business

A CHAPMAN & HALL BOOK

CRC Press
Taylor & Francis Group
6000 Broken Sound Parkway NW, Suite 300
Boca Raton, FL 33487-2742

Printed on acid-free paper
Version Date: 20170823

International Standard Book Number-13: 978-1-4987-6393-6 (Hardback)

Visit the Taylor & Francis Web site at
http://www.taylorandfrancis.com

and the CRC Press Web site at
http://www.crcpress.com

Contents

III   New Challenges and New Solutions for Information Search and Discovery   85

Preface

This is my motto: Analysis is nothing, data are everything. Today, on the web, we can have baskets full of data ... baskets or bins?

Jean-Paul Benzécri, 2011

This book describes solid and supportive foundations for the data science of our times, with many illustrative cases. Core to these foundations are mathematics and computational science. Our thinking and decision-making in regard to data can follow the insightful observation by the physicist Paul Dirac that physical theory and physical meaning have to follow behind the mathematics (see Section 4.7). The hierarchical nature of complex reality is part and parcel of this mathematically well-founded way of observing and interacting with physical, social and all realities.

Quite wide-ranging case studies are used in this book. The text, however, is written in an accessible and easily grasped way, for a reader who is knowledgeable and engaged, without necessarily being an expert in all matters. Ultimately this book seeks to inspire, motivate and orientate our human thinking and acting regarding data, associated information and derived knowledge. This book seeks to give the reader a good start towards practical and meaningful perspectives. Also, by seeking to chart out future perspectives, this book responds to current needs in a way that is unlike other books of some relevance to this field, and that may be great in their own specialisms.

The field of data science has come into its own, in a highly profiled way, in recent times. Ever increasing numbers of employees are required nowadays, as data scientists, in sectors that range from retail to regulatory, and so much besides. Many universities, having started graduate-level courses in data science, are now also starting undergraduate courses. Data science encompasses traditional disciplines of computational science and statistics, data analysis, machine learning and pattern recognition. But new problem domains are arising. Back in the 1970s and into the 1980s, one had to pay a lot of attention to available memory storage when working with computers. Therefore, that focus of attention was on stored data directly linked to the computational processing power. By the beginning of the 1990s, communication and networking had become the focus of attention. Against the background of regulatory and proprietary standards, and open source communication protocols (ISO standards, Decnet, TCP/IP protocols, and so on), data access and display protocols became so central (File Transfer Protocol, gopher, Veronica, Wide Area Information Server, and Hypertext Transfer Protocol). So the focus back in those times was on: firstly, memory and computer power; and secondly, communications and networking. Now we have, thirdly, data as the prime focus. Such waves of technology developments are exciting. They motivate the tackling of new problems, and also there may well be the requirement for new ways of addressing problems. Such requirement of new perspectives and new approaches is always due to contemporary inadequacies, limitations and underperformance. Now, we move on to our interacting with data.

This book targets rigour, and mathematics, and computational thinking. Through available data sets and R code, reproducibility by the reader of results and outcomes is facilitated. Indeed, understanding is also facilitated through "learning by doing". The case studies and the available data and software codes are intended to help impart the data science philosophy in the book. In that sense, dialoguing with data, and "letting the data speak" (Jean-Paul Benzécri), are the perspective and the objective. To the foregoing quotations, the following will be added: "visualization and verbalization of data" (cf. [34]).

Our approach is influenced by how the leading social scientist, Pierre Bourdieu, used the most effective inductive analytics developed by Jean-Paul Benzécri. This family of geometric data analysis methodologies, centrally based on correspondence analysis encompassing hierarchical clustering, and statistical modelling, not only organizes the analysis methodology and domain of application but, most of all, integrates them. An inspirational set of principles for data analytics, listed in [24] (page 6), included the following: "The model should follow the data, and not the reverse. ... What we need is a rigorous method that extracts structures from data." Closely coupled to this is that "data synthesis" could be considered as equally if not more important relative to "data analysis" [27]. Analysis and synthesis of data and information obviously go hand in hand.

A very minor note is the following. Analytics refers to general and generic data processing, obtaining information from data, while analysis refers to specific data processing.

We have then the following. "If I make extensive use of correspondence analysis, in preference to multivariate regression, for instance, it is because correspondence analysis is a relational technique of data analysis whose philosophy corresponds exactly to what, in my view, the reality of the social world is. It is a technique which 'thinks' in terms of relation, as I try to do precisely in terms of field" (Bourdieu, cited in [133, p. 43]).

"In Data Analysis, numerous disciplines need to collaborate. The role of mathematics, although essential, is modest, in the sense that one uses almost exclusively classical theorems or elementary demonstration techniques. But it is necessary that certain abstract conceptions enter into the spirits of the users, the specialists who collect the data and who should orientate the analysis according to fundamental problems that are appropriate to their science" [27].

No method is fruitful unless the data are relevant: "analysing data is not the collecting of disparate data and seeing what comes out of the computer" [27]. In contradistinction to statistics being "technical control" of process, certifying that work has been carried out in conformance with rules, there with primacy accorded to being statistically correct, even asking if such and such a procedure has the right to be used – in contradistinction to that, there is relevance, asking if there is interest in using such and such a procedure.

Another inspirational quotation is that "the construction of clouds leads to the mastery of multidimensionality, by providing 'a tool to make patterns emerge from data'" (this is from Benzécri's 1968 Honolulu conference, when the 1969 proceedings had the paper, "Statistical analysis as a tool to make patterns emerge from data"). John Tukey (developer of exploratory data analysis, i.e. visualization in statistics and data analysis, the fast Fourier transform, and many other methods) expressed this as follows: "Let the data speak for themselves!" This can be kept in mind relative to direct, immediate, unmediated statistical hypothesis testing that relies on a wide range of assumptions (e.g. normality, homoscedasticity, etc.) that are often unrealistic and unverifiable.

The foregoing and the following are in [130]. "Data analysis, or more particularly geometric data analysis is the multivariate statistical approach, developed by J.-P. Benzécri around correspondence analysis, in which data are represented in the form of clouds of points and the interpretation is first and foremost on the clouds of points."

While these are our influences, it would be good, too, to note how new problem areas of Big Data are of concern to us, and also issues of Big Data ethics. A possible ethical issue, entirely due to technical aspects, in the massification and reduction through scale effects that are brought about by Big Data. From [130]: "Rehabilitation of individuals. The context model is always formulated at the individual level, being opposed therefore to modelling at an aggregate level for which the individuals are only an 'error term' of the model."

Now let us look at the importance of homology and field, concepts that are inherent to Bourdieu's work. The comprehensive survey of [108] sets out new contemporary issues of sampling and population distribution estimation. An important take-home message is this: "There is the potential for big data to evaluate or calibrate survey findings ... to help to validate cohort studies". Examples are discussed of "how data ... tracks well with the official", and contextual, repository or holdings. It is well pointed out how one case study discussed "shows the value of using 'big data' to conduct research on surveys (as distinct from survey research)". Therefore, "The new paradigm means it is now possible to digitally capture, semantically reconcile, aggregate, and correlate data."

Limitations, though, are clear [108]: "Although randomization in some form is very beneficial, it is by no means a panacea. Trial participants are commonly very different from the external ... pool, in part because of self-selection". This is because "One type of selection bias is self-selection (which is our focus)".

Important points towards addressing these contemporary issues include the following [108]: "When informing policy, inference to identified reference populations is key". This is part of the bridge which is needed between data analytics technology and deployment of outcomes. "In all situations, modelling is needed to accommodate non-response, dropouts and other forms of missing data."

While "Representativity should be avoided", here is an essential way to address in a fundamental way what we need to address [108]: "Assessment of external validity, i.e. generalization to the population from which the study subjects originated or to other populations, will in principle proceed via formulation of abstract laws of nature similar to physical laws".

The bridge between the data that is analysed, and the calibrating Big Data, is well addressed by the geometry and topology of data. Those form the link between sampled data and the greater cosmos. Pierre Bourdieu's concept of field is a prime exemplar. Consider, as noted in [132], how Bourdieu's work involves "putting his thinking in mathematical terms", and that it "led him to a conscious and systematic move toward a geometric frame-model". This is a multidimensional "structural vision". Bourdieu's analytics "amounted to the global [hence Big Data] effects of a complex structure of interrelationships, which is not reducible to the combination of the multiple [effects] of independent variables". The concept of field, here, uses geometric data analysis that is core to the integrated data and methodology approach used in the correspondence analysis platform [177].

In addressing the "rehabilitation of individuals", which can be considered as addressing representativity both quantitatively as well as qualitatively, there is the potential and relevance for the many ethical issues related to Big Data, detailed in [199]. We may say that in the detailed case study descriptions in that book, what is unethical is the arbitrary representation of an individual by a class or group.

The term *analytics platform* for the science of data, which is quite central to this book, can be associated with an interesting article by *New York Times* author Steve Lohr [146] on the "platform thinking" of the founders of Microsoft, Intel and Apple. In this book the analytics platform is paramount, over and above just analytical or software tools. In his article [146] Lohr says: "In digital-age competition, the long goal is to establish an industry-spanning platform rather than merely products. It is platforms that yield the lucrative flywheel of network effects, complementary products and services and increasing returns." In this book we describe a data analytics platform. It is to have the potential to go way beyond mere tools. It is to be accepted that software tools, incorporating the needed algorithms, can come to one's aid in the nick of time. That is good. But for a deep understanding of all aspects of potential (i.e. having potential for further usage and benefit) and practice, "platform" is the term used here for the following: potential importance and relevance, and a really good conceptional understanding or role. The excellent data analyst does not just come along with a software bag of tricks. The outstanding data analyst will always strive for full integration of theory and practice, of methodology and its implementation.

An approach to drawing benefit from Big Data is precisely as described in [108]. The observation of the need for the "formulation of abstract laws" that bridge sampled data and calibrating Big Data can be addressed, for the data analyst and for the application specialist, as geometric and topological.

In summary, then, this book's key points include the following.

- Our analytics are based on letting the data speak.

- Data synthesis, as well as information and knowledge synthesis, is as important as data analysis.

- In our analytics, an aim too is to rehabilitate the individual (see above).

- We have as a central focus the seeking of, and finding, homology in practice. This is very relevant for Big Data calibration of our analytics.

- In high dimensions, all becomes hierarchical. This is because as dimensionality tends to infinity, and this is a nice perspective on unconscious thought processes, then metric becomes ultametric.

- A major problem of our times may be addressed in both geometric and algebraic ways (remembering Dirac's quotation about the primacy of mathematics even over physics).

- We need to bring new understanding to bear on the dark energy and dark matter of the cosmos that we inhabit, and of the human mind, and of other issues and matters besides. These are among the open problems that haunt humanity.

One major motivation for some of this book's content, related to the fifth item here, is to see, and draw benefit from, the remarkable simplicity of very high dimensions, and even infinite dimensionality. With reference to the last item here, there is a very nice statement by Immanuel Kant, in Chapter 34 of *Critique of Practical Reason* (1788): "Two things fill the mind with ever newer and increasing wonder and awe, the more often and lasting that reflection is concerned with them: the starry sky over me, and the moral law within me."

The Book's Website

The website accompanying this book, which can be found at

http://www.DataScienceGeometryTopology.info

has data sets which are referred to and used in the text. It also has accessible R code which has been used in the many and varied areas of work that are at issue in this book. In some cases, too, there are graphs and figures from outputs obtained.

Provision of data and of some R software, and in a few cases, other software, is with the following objective: to facilitate learning by doing, i.e. carrying out analyses, and reproducing results and outcomes. That may be both interesting and useful, in parallel with the more methodology-related aspects that can be, and that ought to be, revealing and insightful.

Collaborators and Benefactors: Acknowledgements

Key collaborating partners are acknowledged when our joint work is cited throughout the book.

A key stage in this overall work was PhD scholarship funding, with support from the Smith Institute for Industrial Mathematics and System Engineering, and with company support for that, from ThinkingSafe.

Further background were the courses, based on all or very considerable parts of this work, that were taught in April–May 2013 at the First International Conference on Models of Complex Hierarchic Systems and Non-Archimedean Analysis, Cinvestav, Abacus Center, Mexico; and in August 2015 at ESSCaSS 2015, the 14th Estonian Summer School on Computer and Systems Science, Nelijärve Puhkekeskus, Estonia.

Among far-reaching applications of this work there has been a support framework for creative writing that resulted in many books being published. Comparative and qualitative data and information assessment can be well and truly integrated with actionable decision-making. Section 2.7, contains a short overview of these outcomes with quite major educational, publishing and related benefits. It is nice to note that this work was awarded a prestigious teaching prize in 2010, at Royal Holloway University of London. Colleagues Dr Joe Reddington and Dr Douglas Cowie and I, well linked to this book's qualitative and quantitative analytics platform, obtained this award with the title, "Project TooMany-Cooks: applying software design principles to fiction writing".

A number of current collaborations and partnerships, including with corporate and government agencies, will perhaps deliver paradigm-shift advances.

Brief Introduction to Chapters

The chapters of this book are quite largely self-contained, meaning that in a summary way, or sometimes with more detail, there can be essential material that is again presented in any given chapter. This is done so as to take into account the diversity of application domains.

- Chapter 1 relates to the mapping of the semantics, i.e. the inherent meaning and significance of information, underpinning and underlying what is expressed textually and quantitatively. Examples include script story-line analysis, using film script, national research funding, and performance management.

- Chapter 2 relates to a case study of change over time in Twitter. Quantification, including even statistical analysis, of style is motivated by domain-originating stylistic and artistic expertise and insight. Also covered is narrative synthesis and generation.

- Those two chapters comprise Part I, relating to film and movie, literature and documentation, some social media such as Twitter, and the recording, in both quantitative and qualitative ways, of some teamwork activities.

- The accompanying website has as its aim to encourage and to facilitate learning and understanding by doing, i.e. by actively undertaking experimentation and familiarization with all that is described in this book.

- Next comes Part II, relating to underpinning methodology and vantage points.

Paramount are geometry for the mapping of semantics, and, based on this, tree or hierarchical topology, for lots of objectives.

- Chapter 3 relates to how hierarchy can express symmetry. Also at issue is how such symmetries in data and information can be so revealing and informative.

- Chapter 4 is a review chapter, relating to fundamental aspects that are intriguing, and maybe with great potential, in particular for cosmology. This chapter relates to the theme that analytics through real-valued mathematics can be very beneficially complemented by p-adic and, relatedly, m-adic number theory. There is some discussion of relevance and importance in physics and cosmology.

- Part III relates to outcomes from somewhat more computational perspectives.

- Chapter 5 explains the operation of, and the great benefits to be derived from, linear-time hierarchical clustering. Lots of associations with other techniques and so on are included.

- The focus in Chapter 6 is on new application domains such as very high-dimensional data. The chapter describes what we term informally the *remarkable simplicity* of very high-dimensional data, and, quite often, very big data sets and massive data volumes.

- Part IV seeks to describe new perspectives arising out of all of the analytics here, with relevance for various application domains.

- Chapter 7 relates to novel definitions and usage of the concept of *information*.

- Then Chapter 8 relates to ultrametric topology expressing or symbolically representing human unconscious reasoning. Inspiration for this most important and insightful work comes from the eminent psychoanalyst Ignacio Matte Blanco's pursuit of bi-logic, the human's two modes of being, conscious and unconscious.

- Chapter 9 takes such analytics further, with application to very varied expressions of narrative, embracing literature, event and experience reporting.

- Chapter 10 discusses a little the broad and general application of methods at issue here.

Part I

Narratives from Film and Literature, from Social Media and Contemporary Life

1

The Correspondence Analysis Platform for Mapping Semantics

1.1 The Visualization and Verbalization of Data

All-important for the big picture to be presented is introductory description of the geometry of data, and how we can proceed to both visualizing data and interpreting data. We can even claim to be verbalizing our data. To begin with, the embedding of our data in a metric space is our very central interest in the geometry of data. This metric space provides a latent semantic representation of our data. Semantics, or meaning, comes from the sum total of the interrelations of our observations or objects, and of their attributes or properties. Our particular focus is on mathematical description of our data analysis platform (or framework).

We then move from the geometry of metric spaces to the hierarchical topology that allows our data to be structured into clusters.

We address both the mathematical framework and underpinnings, and also algorithms. Hand in hand with the algorithms goes implementation in R (typically).

Contemporary information access is very often *ad hoc*. Querying a search engine addresses some user needs, with content that is here, there and anywhere. Information retrieval results in bits and pieces of information that are provided to the user. On the other hand, information synthesis can refer to the fact that multiple documents and information sources will provide the final and definitive user information. This challenge of Big Data is now looming (J. Mothe, personal communication): "Big Data refers to the fact that data or information is voluminous, varied, and has velocity but above all that it can lead to value provided that its veracity has been properly checked. It implies new information system architecture, new models to represent and analyse heterogeneous information but also new ways of presenting information to the user and of evaluating model effectiveness. Big Data is specifically useful for competitive intelligence activities." It is this outcome that is a good challenge, that is to be addressed through the geometry and topology of data and information: "aggregating information from heterogeneous resources is unsolved."

We can and we will anticipate various ways to address these interesting new challenges. Jean-Paul Benzécri, who was ahead of his time in so many ways, indicated (including in [27]) that "data synthesis" could be considered as equally if not more important relative to "data analysis". Analysis and synthesis of data and information obviously go hand in hand.

Data analytics are just one side of what we are dealing with in this book. The other side, we could say, is that of inductive data analysis. In the context or framework of practical data-related and data-based activity, the processes of data synthesis and inductive data analysis are what we term a narrative. In that sense, we claim to be tracing and tracking the lives of narratives. That is, in physical and behavioural activities, and of course in mental and thought processes.

1.2   Analysis of Narrative from Film and Drama

1.2.1   Introduction

We study two aspects of information semantics: (i) the collection of all relationships; (ii) tracking and spotting anomaly and change. The first is implemented by endowing all relevant information spaces with a Euclidean metric in a common projected space. The second is modelled by an induced ultrametric. A very general way to achieve a Euclidean embedding of different information spaces based on cross-tabulation counts (and from other input data formats) is provided by correspondence analysis. From there, the induced ultrametric that we are particularly interested in takes a sequential (e.g. temporal) – ordering of the data into account. We employ such a perspective to look at narrative, "the flow of thought and the flow of language" [45]. In application to policy decision-making, we show how we can focus analysis in a small number of dimensions.

The data mining and data analysis challenges addressed are the following.

- Great masses of data, textual and otherwise, need to be exploited and decisions need to be made. Correspondence analysis handles multivariate numerical and symbolic data with ease.

- Structures and interrelationships evolve in time.

- We must consider a complex web of relationships.

- We need to address all these issues from data sets and data flows.

Various aspects of how we respond to these challenges will be discussed in this chapter, complemented by the annex to the chapter. We will look at how this works, using the *Casablanca* film script. Then we return to the data mining approach used, to propose that various issues in policy analysis can be addressed by such techniques also.

1.2.2   The Changing Nature of Movie and Drama

McKee [153] bears out the great importance of the film script: "50% of what we understand comes from watching it being said." And: "A screenplay waits for the camera. ... Ninety percent of all verbal expression has no filmic equivalent."

An episode of a television series costs [177] \$2–3 million per hour of television, or £600,000–800,000 for a similar series in the UK. Generally screenplays are written speculatively or commissioned, and then prototyped by the full production of a pilot episode. Increasingly, and especially availed of by the young, television series are delivered via the Internet.

Originating in one medium – cinema, television, game, online – film and drama series are increasingly migrated to another. So scriptwriting must take account of digital multimedia platforms. This has been referred to in computer networking parlance as "multiplay" and in the television media sector as a "360 degree" environment.

Cross-platform delivery motivates interactivity in drama. So-called reality TV has a considerable degree of interactivity, as well as being largely unscripted.

There is a burgeoning need for us to be in a position to model the semantics of film script, – its most revealing structures, patterns and layers. With the drive towards interactivity, we also want to leverage this work towards more general scenario analysis. Potential applications are to business strategy and planning; education and training; and science, technology and economic development policy. We will discuss initial work on the application to policy decision-making in Section 1.3 below.

1.2.3 Correspondence Analysis as a Semantic Analysis Platform

For McKee [153], film script text is the "sensory surface of a work of art" and reflects the underlying emotion or perception. Our data mining approach models and tracks these underlying aspects in the data. Our approach to textual data mining has a range of novel elements.

Firstly, a novelty is our focus on the orientation of narrative through correspondence analysis [24, 171] which maps scenes (and sub-scenes) and words used, in a largely automated way, into a Euclidean space representing all pairwise interrelationships. Such a space is ideal for visualization. Interrelationships between scenes are captured and displayed, as well as interrelationships between words, and mutually between scenes and words. In a given context, comprehensive and exhaustive data, with consequent understanding and use of one's actionable data, are well and truly integrated in this way.

The starting point for analysis is frequency of occurrence data, typically the ordered scenes crossed by all words used in the script.

If the totality of interrelationships is one facet of semantics, then another is anomaly or change as modelled by a clustering hierarchy. If, therefore, a scene is quite different from immediately previous scenes, then it will be incorporated into the hierarchy at a high level. This novel view of hierarchy will be discussed further in Section 1.2.5 below.

We draw on these two vantage points on semantics – viz. totality of interrelationships, and using a hierarchy to express change.

Among further work that is covered in Section 1.2.9 and further in Section 2.5 of Chapter 2 is the following. We can design a Monte Carlo approach to test statistical significance of the given script's patterns and structures as opposed to randomized alternatives (i.e. randomized realizations of the scenes). Alternatively, we examine caesuras and breakpoints in the film script, by taking the Euclidean embedding further and inducing an ultrametric on the sequence of scenes.

1.2.4 *Casablanca* Narrative: Illustrative Analysis

The well-known movie *Casablanca* serves as an example for us. Film scripts, such as for *Casablanca*, are partially structured texts. Each scene has metadata, and the body of the scene contains dialogue and possibly other descriptive data. The *Casablanca* script was half completed when production began in 1942. The dialogue for some scenes was written while shooting was in progress. *Casablanca* was based on an unpublished 1940 screenplay [43]. It was scripted by J.J. Epstein, P.G. Epstein and H. Koch. The film was directed by M. Curtiz and produced by H.B. Wallis and J.L. Warner. It was shot by Warner Bros. between May and August 1942.

As an illustrative first example we use the following. A data set was constructed from the 77 successive scenes crossed by attributes: Int[erior], Ext[erior], Day, Night, Rick, Ilsa, Renault, Strasser, Laszlo, Other (i.e. minor character), and 29 locations. Many locations were met with just once; and Rick's Café was the location of 36 scenes. In scenes based in Rick's Café we did not distinguish between "Main room", "Office", "Balcony", etc. Because of the plethora of scenes other than Rick's Café we assimilate these to just one, "other than Rick's Café", scene.

In Figure 1.1, 12 attributes are displayed. If useful, the 77 scenes can be displayed as dots (to avoid overcrowding of labels). Approximately 34% (for factor 1) + 15% (for factor 2) = 49% of all information, expressed as inertia explained, is displayed here. We can study

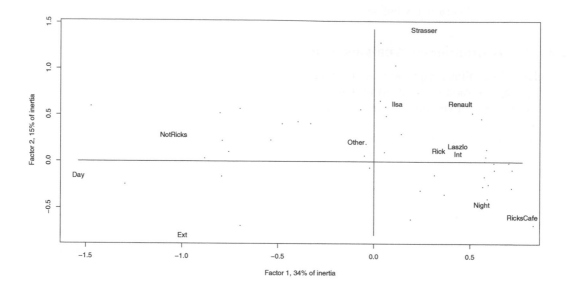

FIGURE 1.1: Correspondence analysis of the *Casablanca* data derived from the script. The input data are presences/absences for 77 scenes crossed by 12 attributes. Just the 12 attributes are displayed. For a short review of the analysis methodology, see the annex to this chapter.

interrelationships between characters, other attributes, and scenes, for instance closeness of Rick's Café with Night and Int (obviously enough).

1.2.5   Modelling Semantics via the Geometry and Topology of Information

Some underlying principles are as follows. We start with the cross-tabulation data, scenes × attributes. Scenes and attributes are embedded in a metric space. This is how we are probing the *geometry of information*, which is a term and viewpoint used by [236].

Underpinning the display in Figure 1.1 is a Euclidean embedding. The triangle inequality holds for metrics. An example of a metric is the Euclidean distance, exemplified in Figure 1.2(a), where each and every triplet of points satisfies the relationship $d(x, z) \leq d(x, y) + d(y, z)$ for distance d. Two other relationships also must hold: symmetry $(d(x, y) = d(y, x))$ and positive definiteness $(d(x, y) > 0$ if $x \neq y$, $d(x, y) = 0$ if $x = y)$.

Further underlying principles used in Figure 1.1 are as follows. The axes are the principal axes of inertia. Principles identical to those in classical mechanics are used. The scenes are located as weighted averages of all associated attributes, and vice versa.

Huyghens' theorem (see Figure 1.2(b)) relates to decomposition of inertia of a cloud of points. This is the basis of correspondence analysis.

We come now to a different principle: that of the *topology of information*. The particular topology used is that of hierarchy. Euclidean embedding provides a very good starting point to look at hierarchical relationships. One particular innovation in this work is as follows: the hierarchy takes sequence (e.g. timeline) into account. This captures, in a more easily understood way, the notions of novelty, anomaly or change.

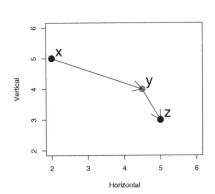

(a) The triangle inequality defines a metric: every triplet of points satisfies the relationship $d(x, z) \leq d(x, y) + d(y, z)$ for distance d.

(b) Christiaan Huyghens, 1629–1695, from [24]. Towards the bottom on the right there is a depiction of the decomposition of the inertia of a hyperellipsoid cloud.

FIGURE 1.2: (a) Depiction of the triangle inequality. Consider a journey from location x to location z, but via y. (b) A poetic portrayal of Huyghens.

Let us take an informal case study to see how this works. Consider the situation of seeking documents based on titles. If the target population has at least one document that is close to the query, then this is (let us assume) clear-cut. However, if all documents in the target population are very unlike the query, does it make any sense to choose the closest? Whatever the answer, here we are focusing on the inherent ambiguity, which we will note or record in an appropriate way. Figure 1.3(a) illustrates this situation, where the query is the point to the right. By using approximate similarity the situation can be modelled as an isosceles triangle with small base.

As illustrated in Figure 1.3(a), we are close to having an isosceles triangle with small base, with the red dot as apex, and with a pair of the black dots as the base. In practice, in hierarchical clustering, we fit a hierarchy to our data. An ultrametric space has properties that are very unlike a metric space, and one such property is that the only triangles allowed are either equilateral, or isosceles with small base. So Figure 1.3(a) can be taken as representing a case of ultrametricity. What this means is that the query can be viewed as having a particular sort of dominance or hierarchical relationship *vis-à-vis* any pair of target documents. Hence any triplet of points here, one of which is the query (defining the apex of the isosceles, with small base, triangle), defines local hierarchical or ultrametric structure. Further general discussion can be found in [169], including how established nearest neighbour or best match search algorithms often employ such principles.

It is clear from Figure 1.3(a) that we should use approximate equality of the long sides of the triangle. The further away the query is from the other data, the better is this approximation [169].

What sort of explanation does this provide for our example here? It means that the query is a novel, or anomalous, or unusual "document". It is up to us to decide how to treat

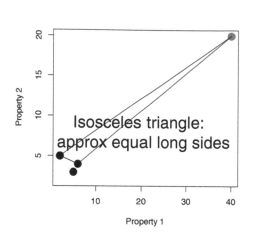

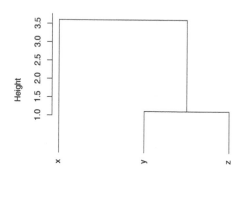

(a) The query is on the upper right. While we can easily determine the closest target (among the three objects represented by the dots on the left), is the closest really that much different from the alternatives?

(b) The strong triangle inequality defines an ultrametric: every triplet of points satisfies the relationship $d(x,z) \leq \max\{d(x,y), d(y,z)\}$ for distance d. Check by reading off the hierarchy, how this is verified for all x, y, z: $d(x,z) = 3.5$, $d(x,y) = 3.5$, $d(y,z) = 1.0$. In addition, the symmetry and positive definiteness conditions hold for any pair of points.

FIGURE 1.3: (a) graphical depiction, and (b) hierarchy, or rooted tree, depiction.

such new, innovative cases. It raises, though, the interesting perspective that here we have a way to model and subsequently handle the semantics of anomaly or innocuousness.

The strong triangle inequality, or ultrametric inequality, holds for tree distances: see Figure 1.3(b). The closest common ancestor distance is such an ultrametric.

1.2.6 *Casablanca* Narrative: Illustrative Analysis Continued

Figure 1.4 uses a sequence-constrained complete link agglomerative algorithm. It shows up scenes 9 to 10, and progressing from 39 to 40 and 41, as major changes. The sequence- or chronology-constrained algorithm (i.e. agglomerations are permitted between adjacent segments of scenes only) is described in the annex to this chapter, and in greater detail in [167, 19, 135]. The agglomerative criterion used, that is subject to this sequence constraint, is a complete link one.

1.2.7 Platform for Analysis of Semantics

Correspondence analysis supports the following:

- analysis of multivariate, mixed numerical/symbolic data;
- web (viz. pairwise links) of interrelationships;
- evolution of relationships over time.

Correspondence analysis is in practice *a tale of three metrics* [171]. The analysis is based

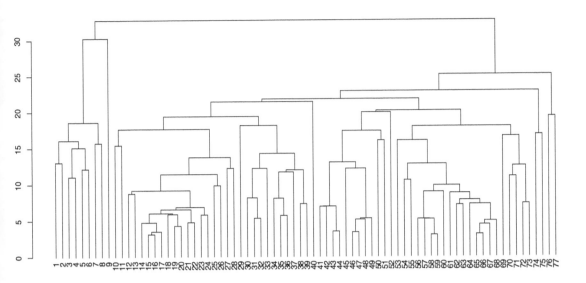

FIGURE 1.4: The 77 scenes clustered. These scenes are in sequence: a sequence-constrained agglomerative criterion is used for this. The agglomerative criterion itself is a complete link one. See [167] for properties of this algorithm.

on embedding a cloud of points from a space governed by one metric into another. The cloud of observables is inherently related to the cloud of attributes of those observables. Observables are defined by their attributes, and each attribute is, *de facto*, specified by its associated observables. So – in the case of film script – for any one of the metrics we can effortlessly pass between the space of film script scenes and attribute set. The three metrics are as follows.

- Chi-squared (χ^2) metric, appropriate for profiles of frequencies of occurrence.

- Euclidean metric, for visualization, and for static context.

- Ultrametric, for hierarchic relations and for dynamic context, as we operationally have it here, also taking the chronology into account.

In the analysis of semantics, we distinguish two separate aspects.

1. Context – the collection of all interrelationships.

    - The Euclidean distance makes a lot of sense when the population is homogeneous.
    - All interrelationships together provide context, relativities – and hence meaning.

2. Hierarchy tracks anomaly.

    - Ultrametric distance makes a lot of sense when the observables are heterogeneous, discontinuous.
    - The latter is especially useful for determining anomalous, atypical, innovative cases.

1.2.8   Deeper Look at Semantics of *Casablanca*: Text Mining

The *Casablanca* script has 77 successive scenes. In total there are 6710 words in these scenes. We define words here as consisting of at least two letters. Punctuation is first removed. All upper case is set to lower case. We analyse frequencies of occurrence of words in scenes, so the input is a matrix crossing scenes by words.

1.2.9   Analysis of a Pivotal Scene

As a basis for a deeper look at *Casablanca* we have taken comprehensive but qualitative discussion by McKee [153] and sought quantitative and algorithmic implementation.

Casablanca is based on a range of miniplots. For McKee its composition is "virtually perfect".

Following McKee [153], we will carry out an analysis of *Casablanca*'s "mid-act climax", scene 43. McKee divides this scene, relating to Ilsa and Rick seeking black market exit visas, into 11 "beats".

1. Beat 1 is Rick finding Ilsa in the market.

2. Beats 2, 3, 4 are rejections of him by Ilsa.

3. Beats 5, 6 express rapprochement by both.

4. Beat 7 is guilt-tripping by each in turn.

5. Beat 8 is a jump in content: Ilsa says she will leave Casablanca soon.

6. In beat 9, Rick calls her a coward, and Ilsa calls him a fool.

7. In beat 10, Rick propositions her.

8. In beat 11, the climax, all goes to rack and ruin: Ilsa says she was married to Laszlo all along. Rick is stunned.

Figure 1.5 shows the evolution from beat to beat rather well. In these 11 beats or subscenes 210 words are used. Beat 8 is a dramatic development. Moving upwards on the ordinate (factor 2) indicates distance between Rick and Ilsa. Moving downwards indicates rapprochement.

In the full-dimensional space we can check some other of McKee's guidelines. Lengths of beat get shorter, leading up to climax: word counts of the final five beats in scene 43 are: 50, 44, 38, 30, 46. A style analysis of scene 43 based on McKee [153] can be Monte Carlo tested against 999 uniformly randomized sets of the beats. In the great majority of cases (against 83% and more of the randomized alternatives) we find the style in scene 43 to be characterized by: small variability of movement from one beat to the next; greater tempo of beats; and high mean rhythm. There is further description of these attributes in Section 2.5.

The planar representation in Figure 1.5 accounts for 12.6% + 12.2% = 24.8% of the inertia, and hence the total information. We will look at the evolution of scene 43, using hierarchical clustering of the full-dimensional data – but based on the relative orientations, or correlations with factors. This is because of what we have found in Figure 1.5, namely that *change* of direction is most important.

Figure 1.6 shows the hierarchical clustering, based on the sequence of beats. Input data are of full dimensionality so there is no approximation involved. Note the caesura in moving from beat 7 to 8, and back to 9. There is less of a caesura in moving from 4 to 5, but it is still quite pronounced.

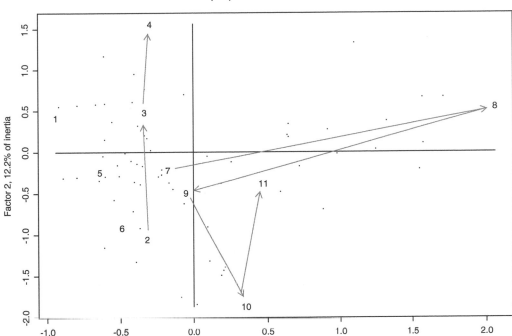

FIGURE 1.5: Correspondence analysis principal plane – best Euclidean embedding in two dimensions – of scene 43. This scene is a central and indeed a pivotal one in *Casablanca*. It consists of eleven sub-scenes, which McKee terms "beats". Discussed in the text is the evolution over sub-scenes 2–4; and again over sub-scenes 7–11.

1.3   Application of Narrative Analysis to Science and Engineering Research

Our way of analysing semantics is sketched out as follows:

• We discern story semantics arising out of the orientation of narrative.

• This is based on the web of interrelationships.

• We examine caesuras and breakpoints in the flow of narrative.

Let us look at the implications of this for data mining with decision policy support in view.

Consider a fairly typical funded research project, and its phases up to and beyond the funding decision. Different research funding agencies differ in their procedures. But a narrative can always be strung together. All stages of the proposal and successful project life cycle, including external evaluation and internal decision-making, are highly document – and as a consequence narrative – based. Putting all phases together we have a story-line, which provides in effect a narrative.

As a first step towards much fuller analysis of many and varied narratives involved in

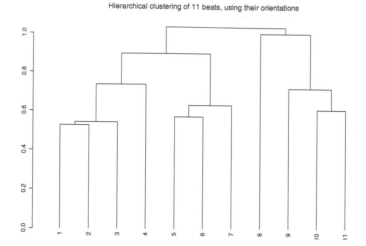

FIGURE 1.6: Hierarchical clustering of sequence of beats in scene 43 of *Casablanca*. Again, a sequence-constrained complete link agglomerative clustering algorithm is used. The input data is based on the full-dimensionality Euclidean embedding provided by the correspondence analysis.

research, and research and development (R&D) funded projects, let us look at the very general role of narrative in national research development. We look here at:

- overall view – overall synthesis of information;
- orientation of strands of development;
- their tempo, rhythm.

Through such an analysis of narrative, among the follow-on implications for further analytics to be addressed are:

- strategy and its implementation in terms of themes and sub-themes represented;
- thematic focus and coverage;
- organizational clustering;
- evaluation of outputs in a global context;
- all the above over time.

The aim is to understand the "big picture". It is not to replace the varied measures of success that are applied, such as publications, patents, licences, numbers of PhDs completed, company start-ups, and so on. It is instead to appreciate the broader configuration and orientation, and to determine the most salient aspects underlying the data.

1.3.1　Assessing Coverage and Completeness

When I was managing national research funding, the following were the largest funded units: Science Foundation Ireland (SFI) Centres for Science, Engineering and Technology

(CSETs), campus–industry partnerships typically funded at up to € 20 million over 5 years; Strategic Research Clusters (SRCs), also research consortia, with industrial partners and over 5 years typically funded at up to € 7.5 million.

We cross-tabulated eight CSETs and 12 SRCs by a range of terms derived from title and summary information, together with budget, numbers of principal investigators (PIs), co-investigators (Co-Is), and PhDs. We can display any or all of this information on a common map, for visual convenience a planar display, using correspondence analysis.

In mapping SFI CSETs and SRCs, now correspondence analysis is employed, based on the upper (near root) part of an ontology or concept hierarchy. This we propose as *information focusing*. Correspondence analysis provides simultaneous representation of observations and attributes. Retrospectively, we can project other observations or attributes into the factor space: these are supplementary observations or attributes. A two-dimensional or planar view is likely to be a gross approximation of the full cloud of observations or of attributes. We may accept such an approximation as insightful and informative. Another way to address this same issue is as follows. We define a small number of aggregates of either observations or attributes, and carry out the analysis on them. We then project the full set of observations and attributes into the factor space. For mapping of SFI CSETs and SRCs a simple algebra of themes as set out in the next paragraph achieves this goal. The upshot is that the two-dimensional or planar view provides a very good fit to the full cloud of observations or of attributes.

From CSET or SRC characterization as: Physical Systems (Phys), Logical Systems (Log), Body/Individual, Health/Collective, and Data & Information (Data), the following thematic areas were defined:

1. eSciences = Logical Systems, Data & Information

2. Biosciences = Body/Individual, Health/Collective

3. Medical = Body/Individual, Health/Collective, Physical Systems

4. ICT = Physical Systems, Logical Systems, Data & Information

5. eMedical = Body/Individual, Health/Collective, Logical Systems

6. eBiosciences = Body/Individual, Health/Collective, Data & Information.

This categorization scheme can be viewed as the upper level of a concept hierarchy. It can be contrasted with the somewhat more detailed scheme that we used for analysis of published journal articles. The author's *Computer Journal* editorial [174] described this.

CSETs labelled in the figures are: APC, Alimentary Pharmabiotic Centre; BDI, Biomedical Diagnostics Institute; CRANN, Centre for Research on Adaptive Nanostructures and Nanodevices; CTVR, Centre for Telecommunications Value-Chain Research; DERI, Digital Enterprise Research Institute; LERO, Irish Software Engineering Research Centre; NGL, Centre for Next Generation Localization; and REMEDI, Regenerative Medicine Institute.

In Figure 1.7 eight CSETs and major themes are shown. Factor 1 counterposes computer engineering (left) to biosciences (right). Factor 2 counterposes software on the positive end to hardware on the negative end. This two-dimensional map encapsulates 64% (for factor 1) + 29% (for factor 2) = 93% of all information (i.e. inertia) in the dual clouds of points. CSETs are positioned relative to the thematic areas used. In Figure 1.8 sub-themes are additionally projected into the display. This is done by taking the sub-themes as *supplementary elements* following the analysis as such: see the annex to this chapter for a short introduction to this. From Figure 1.8 we might wish to label additionally factor 2 as a polarity of data and physics, associated with the extremes of software and hardware.

In Figure 1.9 CSET budgets are shown, in millions of euros over 5 years, and themes

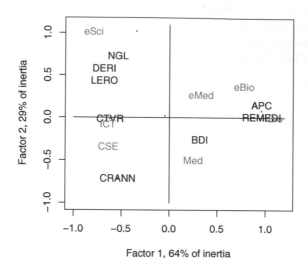

FIGURE 1.7: CSETs, labelled, with themes located on a planar display, which is nearly complete in terms of information content.

are also displayed. In this way we use the map to show characteristics of the CSETs, in this case budgets.

Figure 1.10 shows 12 SRCs that started at the end of 2007. The planar space into which the SRCs are projected is identical to Figures 1.7–1.9. This projection is accomplished by supplementary elements (see the annex to this chapter).

Figure 1.11 shows one property of the SRCs, their budgets in millions of euros over 5 years.

1.3.2   Change over Time

We take another funding programme, the Research Frontiers Programme, to show how changes over time can be mapped.

This annual funding programme included all fields of science, mathematics and engineering. There were approximately 750 submissions annually, with 168 funding awards in 2007, of average size € 155,000, and 143 funding awards in 2008, of average size € 161,000, for these 3–4-year research projects. We will look at the Computer Science panel results only, over 2005, 2006, 2007 and 2008.

Grants awarded in these years were respectively 14, 11, 15, 17. The breakdown by universities and other third-level higher education institutes concerned was: UCD, 13; TCD, 10; DCU, 14; UCC, 6; UL, 3; DIT, 3; NUIM, 3; WIT, 1.

One theme was used to characterize each proposal from among the following: bioinformatics, imaging/video, software, networks, data processing & information retrieval, speech and language processing, virtual spaces, language ND text, information security, and e-learning. Again this categorization of computer science can be contrasted with one used for articles in the *Computer Journal* [174].

Figures 1.12–1.14 show different facets of the Computer Science outcomes. By keeping the displays separate, we focus on one aspect at a time. All displays, however, are based

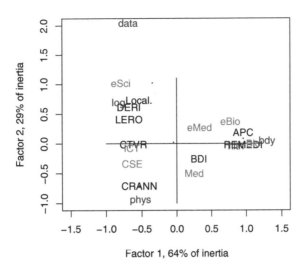

FIGURE 1.8: As Figure 1.7 but with sub-themes projected into the display. Note that, through use of supplementary elements, the axes and scales are identical to Figures 1.7, 1.9 and 1.10. Axes and scales are just displayed differently in this figure so that sub-themes appear in our field of view.

on the same list of themes, and so allow mutual comparisons. Note that the principal plane shown accounts for $9.5\% + 8.9\% = 18.4\%$ of the overall inertia, that in turn expresses information here. Although small, it is the best planar view of the data (arising from the chi-squared metric (see the annex to this chapter), followed by the Euclidean embedding that the figures show). Ten themes were used, and what the 18.4% information content tells us is that there is importance attached to most if not all of the ten. We are not prevented, though, from usefully studying the planar displays. That they can be used to display lots of supplementary data is a major benefit of their use.

What the analyses demonstrate is that the categories used are of crucial importance. Indeed, in Figures 1.7–1.11 and then in Figures 1.12–1.14, we see how we can "engineer" the impact of the categories by assimilating their importance to moments of inertia of the clouds of associated points.

1.3.3   Conclusion on the Policy Case Studies

The aims and objectives in our use of the correspondence analysis and clustering platform is to drive strategy and its implementation in policy.

What we are targeting is to study highly multivariate, evolving data flows. This is in terms of the semantics of the data – principally, complex webs of interrelationships and evolution of relationships over time. This is the *narrative of process* that lies behind raw statistics and funding decisions.

We have been concerned especially with *information focusing* in Section 1.3.1, and this over time in Section 1.3.2.

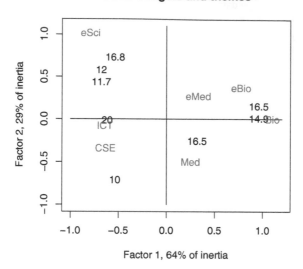

FIGURE 1.9: Similar to Figure 1.7, but here we show CSET budgets.

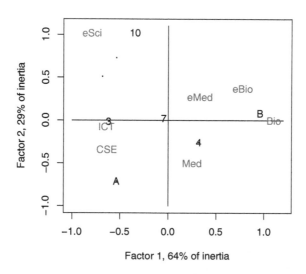

FIGURE 1.10: Using the same themes, the SRCs are projected. The properties of the planar display are the same as for Figures 1.7–1.9. SRCs 3, 4, 7, 10 are shown; and overlapping groups of four each are at A and B. A represents four bioscience or pharmaceutical SRCs. B represents four materials, biomaterials, or photonics SRCs.

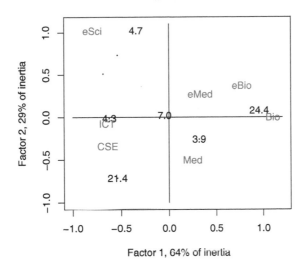

FIGURE 1.11: As Figure 1.10, now displaying combined budgets.

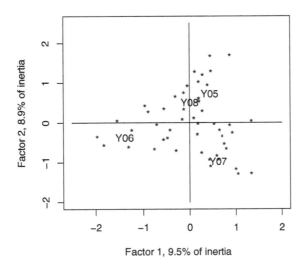

FIGURE 1.12: Research Frontiers Programme over 4 years. Successful proposals are shown as asterisks. The years are located as the average of successful projects.

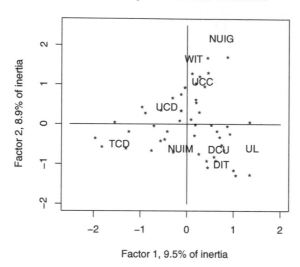

FIGURE 1.13: As Figure 1.12, displaying host institutes of the awardees.

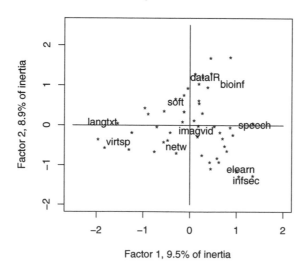

FIGURE 1.14: As Figures 1.12 and 1.13, displaying themes.

Staff	Ext	Int	V3 Evg	Gra				
1	3	2	2	2	0	1	1	1
2	2	3	2	2.5	1	0	1	0.5
3	3	3	3	2.5	0	0	0	0.5
4	3	3	3	3	0	0	0	0
5	3	2.5	2	2	0	0.5	1	1
6	2	2	2	2	1	1	1	1
7	3	3	3	3	0	0	0.5	0.5
8	3	3	2.5	2.5	0	0	0.5	0.5
9	3	2.5	2.5	2.5	0	0.5	0.5	0.5
10	3	2.5	2.5	3	0	0.5	0.5	0

TABLE 1.1: On the left is the original set of scores for 10 staff members on the 4 criteria. Then on the right is the complement of these scores. The right-hand side shows the lacking scores, what the author, as manager, deemed to be lacking.

1.4 Human Resources Multivariate Performance Grading

In this section, the objective is performance mapping. Apart from time changes that are easily displayed, we seek to account fully for, and indeed assess, the evaluation criteria that are used. The following small example is of an annual appraisal exercise of 10 staff members who were managed by the author. The multivariate nature of corporate performance was defined in the following way. Each individual staff member was scored 1 (adequate), 2 (high level) or 3 (above expectation) in terms of demonstrated performance along the following four orientations.

1. External professionalism: authority and clarity in delivery in regard to all tasks assigned. Includes: routine, standard procedures; and particular issues and problems arising.

2. Internal professionalism: reporting on all (relevant) levels.

3. Advocacy and, where relevant, evangelism, in regard to socio-technological sectors that are powered or empowered by outcomes or offshoots of our work.

4. Personal positioning *vis-à-vis* role when graduated from the organization, i.e. when, having left the organization in the future, there would be both gained expertise and possible, mutually beneficial, positive continuing and extended profile-building.

The following is a small note regarding the last criterion here, used by the author in his role as director. That is, that an employee leaving the workforce is graduating from the workplace. This may even on some occasions be considered like a student who will graduate.

All of the doubled data table in Table 1.1 is used in the analysis. Doubling of data is when the value and the complement of that value relative to a total score are both used. As a result the row totals are constant, and therefore the row masses in the correspondence analysis will be identical.

Considerations for closer detailing of performance were as follows. Roles can well change from one year to the next (at least), so direction and effectiveness need to be both planned and then later mapped. No job is for life, although corporate memory and culture are lodged with the individual and are not separable from the individual. No one among those assessed

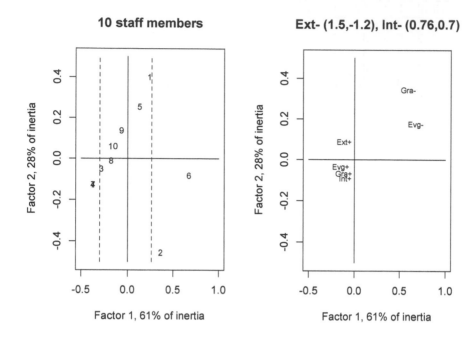

FIGURE 1.15: Principal factor plane accounting for just 90% of the inertia. Staff members 4 and 7 are at identical locations in this planar projection. The dotted lines show the 20th and 80th percentiles. Here, staffers 4, 7 are in the top 20th percentile; and staffers 2, 6 are in the bottom 20th percentile. Off-scale on the display here are indicated, with coordinates in the header, the lacking external facing (Ext−), and internal facing (Int−) performances.

can be said to be replaceable by a better person. That is a finding resulting from the scoring in this exercise.

From Figure 1.15, the achieved performance criteria are close together, defining the common standard of the staff team. That is close to the origin, expressing the average staff profile. In the negative first factor direction there is the best performance, that of overlaid staffers 4 and 7. For staffers 6 and 2, to name a criterion which would be most appropriate to work on, we respond with Ext−, that is, the external professionalism. All lacking attributes (of the doubled data) are towards the positive first factor half-axis.

While we could well display the movement from the previous year's annual appraisal, here we simply note that, compared to our assessment a year before, staffer 7 moved out to be in same position as 4; 10 is a little toward the centre of this factor (the average or standard); and 5 has moved from being in the same position as 2 to being more effective.

The dotted lines in the figure display the top 20th and bottom 20th percentiles. Factor 1 can be looked at in terms of a histogram of performances provided by the projections of staffers on this first factor. That is a long-tailed histogram, expressing the bunching of projections on factor 1 that is clear from staffers 4, 7 onwards (from left to right) in Figure 1.15. Take the histogram on 0.5-length intervals from −0.5 to 1, and we get interval bins with numbers of staffers: 6, 3, 1.

While a normal distribution is often used as the basic assumption in regard to human resources performance, with $\pm 3\sigma$ accounting for 99.7% of variation, in fact, in performance just as in other domains, such as text analysis as will be seen in later chapters, it is nearly always a case of long-tailed, exponentially decreasing distributions. Our analytics platform takes due account of the need to cater for such distributions. In this regard the correspondence analysis platform is more robust, more adaptable and more powerful than simple, straightforward principal components analysis, which caters well for normal or Gaussian clouds.

1.5   Data Analytics as the Narrative of the Analysis Processing

A conclusion now drawn is how data processing in analysis is likened to the narrative of the analysis. In a sense, we are turning the camera back on our own thinking and reflection that takes place when we are carrying out our analytics work.

Data collected on processes or configurations can tell us a lot about what is going on. When backed up with probabilistic models, we can test one hypothesis against another. This leads to data mining "in the small". Pairwise plots are often sufficient because as visualizations they remain close and hence faithful to the original data. Such data mining "in the small" is well served by statistics as a discipline.

Separately from such data mining "in the small", there is a need for data mining "in the large". We need to unravel the broad brush strokes of narrative expressing a story. Indeed, as discussed in [45], narrative expresses not just a story but human thinking. A broad brush view of science and engineering research, and their linkages with the economy, is a similar type of situation. Pairwise plots are not always sufficient now, because they tie us down in unnecessary detail. We use them since we need the more important patterns, trends, details to be brought out in order of priority. Strategy is of the same class as narrative.

For data mining "in the large" the choice of data and categorization used are crucial. This we have exemplified well in this chapter. One consequence is that the factor space mapping may well be nearly fully represented in the principal factor plane. So our data mining "in the large" is tightly coupled to the data. This is how we can make data speak. The modelling involved in analysis "in the small" comes later.

1.6   Annex: The Correspondence Analysis and Hierarchical Clustering Platform

This annex introduces important aspects of correspondence analysis and hierarchical clustering. For further reading, see [25, 24, 171].

1.6.1   Analysis Chain

1. The starting point is a matrix that cross-tabulates the dependencies (e.g. frequencies of joint occurrence) of an observations crossed by attributes matrix.

2. By endowing the cross-tabulation matrix with the χ^2 metric on both observation set (rows) and attribute set (columns), we can map observations and attributes into the same space, endowed with the Euclidean metric.

3. A hierarchical clustering is induced on the Euclidean space, the factor space.

4. Interpretation is through projections of observations, attributes or clusters onto factors. The factors are ordered by decreasing importance.

Various aspects of correspondence analysis follow on from this, such as multiple correspondence analysis, different ways that one can encode input data, and mutual description of clusters in terms of factors and vice versa. In the following we use elements of the Einstein tensor notation of [24]. This often reduces to common vector notation.

1.6.2 Correspondence Analysis: Mapping χ^2 Distances into Euclidean Distances

- The given contingency table (or numbers of occurrence) data is denoted $k_{IJ} = \{k_{IJ}(i,j) = k(i,j); \ i \in I, j \in J\}$.

- I is the set of observation indices, and J is the set of attribute indices. We have $k(i) = \sum_{j \in J} k(i,j)$. We define $k(j)$ analogously, and $k = \sum_{i \in I, j \in J} k(i,j)$.

- Relative frequencies are $f_{IJ} = \{f_{ij} = k(i,j)/k; \ i \in I, j \in J\} \subset \mathbb{R}_{I \times J}$; similarly, f_I is defined as $\{f_i = k(i)/k; \ i \in I\} \subset \mathbb{R}_I$, and f_J analogously.

- The conditional distribution of f_J knowing $i \in I$, also termed the jth *profile* with coordinates indexed by the elements of I, is

$$f_J^i = \{f_j^i = f_{ij}/f_i = (k_{ij}/k)/(k_i/k); \ f_i > 0; \ j \in J\}$$

and likewise for f_I^j.

- What is discussed in terms of information focusing in the text is underpinned by the *principle of distributional equivalence*. This means that if two or more identical profiles are aggregated by simple elementwise summation, then the χ^2 distances relating to other profiles are not affected.

On this last point relating to the principle of distributional equivalence, the essential issue is that aggregating, simply by summing, two identical profiles (i.e. row profiles or column profiles) results in no change for all of the mutual distances. Furthermore, this results in the aggregating of similar profiles changing very little all of the mutual distances. See [28, Section 4.2].

1.6.3 Input: Cloud of Points Endowed with the Chi-Squared Metric

- The cloud of points consists of the couples: (multidimensional) profile coordinate and (scalar) mass. We have $N_J(I) = \{(f_J^i, f_i); \ i \in I\} \subset \mathbb{R}_J$, and again similarly for $N_I(J)$.

- Included in this expression is the fact that the cloud of observations, $N_J(I)$, is a subset of the real space of dimensionality $|J|$ where $|\cdot|$ denotes the cardinality of the attribute set, J.

- The overall inertia is

$$M^2(N_J(I)) = M^2(N_I(J)) = \|f_{IJ} - f_I f_J\|_{f_I f_J}^2 = \sum_{i \in I, j \in J} (f_{ij} - f_i f_j)^2 / f_i f_j.$$

- The term $\|f_{IJ} - f_I f_J\|_{f_I f_J}^2$ is the χ^2 metric between the probability distribution f_{IJ} and the product of marginal distributions $f_I f_J$, with as centre of the metric the product $f_I f_J$. Probability is empirically defined from the frequencies.

- Decomposing the moment of inertia of the cloud $N_J(I)$ – or of $N_I(J)$ since both analyses are inherently related – furnishes the principal axes of inertia, defined from a singular value decomposition.

1.6.4 Output: Cloud of Points Endowed with the Euclidean Metric in Factor Space

- The χ^2 distance with centre f_J between observations i and i' is written as follows in two different notations:

$$d(i, i') = \|f_J^i - f_J^{i'}\|_{f_J}^2 = \sum_j \frac{1}{f_j} \left(\frac{f_{ij}}{f_i} - \frac{f_{i'j}}{f_{i'}} \right)^2.$$

- In the factor space this pairwise distance is identical. There is invariance of the distance, while the coordinate system and the metric change.

- For factors indexed by α and for total dimensionality N ($N = \min\{|I| - 1, |J| - 1\}$; the subtraction of 1 is since the χ^2 distance is centred and hence there is a linear dependency which reduces the inherent dimensionality by 1), we have the projection of observation i on the αth factor, F_α, given by $F_\alpha(i)$:

$$d(i, i') = \sum_{\alpha=1,\ldots,N} (F_\alpha(i) - F_\alpha(i'))^2. \tag{1.1}$$

Analogously for attribute, or modality, j, the αth factor projection is $G_\alpha(j)$.

- In correspondence analysis the factors are ordered by decreasing moments of inertia.

- The factors are closely related, mathematically, in the decomposition of the overall cloud, $N_J(I)$ and $N_I(J)$, inertias.

- The eigenvalues associated with the factors, identically in the space of observations indexed by set I and in the space of attributes indexed by set J, are given by the eigenvalues associated with the decomposition of the inertia.

- The decomposition of the inertia is a principal axis decomposition, which is arrived at through a singular value decomposition.

1.6.5 Supplementary Elements: Information Space Fusion

Dual Spaces and Transition Formulas

- The factor projections are given by

$$F_\alpha(i) = \lambda_\alpha^{-\frac{1}{2}} \sum_{j \in J} f_j^i G_\alpha(j), \quad \text{for } \alpha = 1, 2, \ldots, N; \; i \in I, \tag{1.2}$$

$$G_\alpha(j) = \lambda_\alpha^{-\frac{1}{2}} \sum_{i \in I} f_i^j F_\alpha(i) \quad \text{for } \alpha = 1, 2, \ldots, N; \; j \in J. \tag{1.3}$$

- *Transition formulas*: The coordinate of element $i \in I$ is the barycentre of the coordinates of the elements $j \in J$, with associated masses of value given by the coordinates of f_j^i of the profile f_J^i. This is all to within the $\lambda_\alpha^{-\frac{1}{2}}$ constant.

- In the output display, the barycentric principle comes into play: this allows us to simultaneously view and interpret observations and attributes.

Supplementary Elements

Overly preponderant elements (i.e. row or column profiles) or exceptional elements (e.g. a gender attribute, given other performance or behavioural attributes) may be placed as supplementary elements. Their projections are determined using the transition formulas, equations (1.2) or (1.3), used with f_j^i or f_i^j. This amounts to carrying out a correspondence analysis first, without these elements, and then projecting them into the factor space following the determination of all properties of this space.

Here too we have a new approach to fusion of information spaces focusing the projection.

1.6.6   Hierarchical Clustering: Sequence-Constrained

Background on hierarchical clustering in general, and the particular algorithm used here, can be found in [167].

Consider the projection of observation i onto the set of all factors indexed by α, $\{F_\alpha(i)\}$ for all α, which defines the observation i in the new coordinate frame. This new factor space is endowed with the (unweighted) Euclidean distance, d. We seek a hierarchical clustering that takes into account the observation sequence, that is, observation i precedes observation i' for all $i, i' \in I$. We use the linear order on the observation set. The sequence or adjacency constraint is, in practice for most of what is at issue here, a chronological constraint.

Agglomerative hierarchical clustering algorithm:

1. Consider each observation in the sequence as constituting a singleton cluster. Determine the closest pair of adjacent observations, and define a cluster from them.

2. Determine and merge the closest pair of adjacent clusters, c_1 and c_2, where closeness is defined by $d(c_1, c_2) = \max \{d_{ii'}$ such that $i \in c_1, \ i' \in c_2\}$.

3. Repeat the second step until only one cluster remains.

This is a sequence-constrained complete link agglomeration criterion. The cluster proximity at each agglomeration is strictly non-decreasing. Proof of this, and use of the constrained hierarchical clustering, is further elaborated on, with proofs and applications, in [167, 19, 135].

2

Analysis and Synthesis of Narrative: Semantics of Interactivity

2.1 Impact and Effect in Narrative: A Shock Occurrence in Social Media

Before progressing to stylistic feature analysis, we first give our attention to exceptional or anomalous narrative events. While the scope of this case study, its context, might be considered quite entertaining, nonetheless it is a case study that could be drawn on for deeper and more profound situations, for example relating to security and forensics.

When, in October 2009, the actor, presenter and celebrity Stephen Fry announced his retirement from Twitter to his near 1 million followers, it was a newsworthy event. It was reported [205] that: "Fry's disagreement with another tweeter began when the latter said 'I admire and adore' Fry, but that he found his tweets 'a bit ... boring ... (sorry Stephen)'.

"The tweeter, who said that he had been blocked from viewing Fry's Twitter feed, later apologised and acknowledged that Fry suffers from bipolar disorder."

Having caused major impact among his followers and wider afield, Fry actually returned to Twitter, nearly immediately, having had an apology from the offending tweeter, @brumplum.

Stephen Fry's two crucial tweets were as follows. In description below, they are referred to as, respectively, the "I retire" tweet and the "aggression" tweet. First, at 6:09 a.m. on 31 October 2009:

> @brumplum You've convinced me. I'm obviously not good enough. I retire from Twitter henceforward. Bye everyone.

Then, at 6.13 a.m.:

> Think I may have to give up on Twitter. Too much aggression and unkindness around. Pity. Well, it's been fun.

In order to look at those decisive tweets in context, a set of 302 of Fry's tweets were taken, spanning the critical early morning of 31 October 2009, from 22 October to 22 November.

2.1.1 Analysis

Words are collected from the 302 tweets. Initially we have 1787 unique words defined as follows: containing more than one consecutive letter; with punctuation and special characters deleted (hence with modification of short URLs, hashtags or Twitter names preceded by an @ symbol, but for our purposes not detracting from our interest in semantic content); and with no lemmatization nor other processing, in order to leave us with all available emotionally laden function words. For our analysis we do require a certain amount of sharing of words by the tweets. Otherwise there will be isolated tweets (that are disconnected through no shared terms). So we select words depending on two thresholds: a minimum global frequency and a minimum number of tweets for which the word is used. Both thresholds were set to 5 (determined as a compromise between a good overlap between tweets in terms of word presence, and yet not removing an overly large number of words). This led to 143 words retained for the 302 tweets. A repercussion is that some tweets became empty of words: 293 were non-empty, out of the 302 tweet set.

Since the word usage in each tweet is limited, this results in the Euclidean, latent space factor embedding accounting for limited numbers of tweets and of words. Quite likely – and this was verified from our tweet stream data – just one word and/or just one tweet contributed to most of the inertia of the factor. Thus, such a word and/or such a tweet determined the factor. The cumulative percentages of inertia explained by the succession of factors (i.e. the cumulative percentage relative eigenvalues) were found to be 2.4, 4.5, 6.5, 8.5, 10.4, 12.2, 14.1, 15.9, and onwards finally to 100. This correspondence analysis used a word set (often termed a corpus) of 121 words, having removed a list of stopwords that are listed below, and had 280 non-empty tweets.

Given the lack of any great concentration of inertia in the succession of factors, we developed an analysis methodology as outlined in the following subsections. For high-dimensional word usage spaces, it is to be noted of course that it is normal for correspondence analysis to have such a lack of concentration of inertia in the succession of factors, and hence the factors are close to being equally important (quantified by the inertia about them of the cloud of tweets, or the cloud of words).

2.1.2   Two Critical Tweets in Terms of Their Words

First we pursued the following analysis approach. Taking the two crucial tweets noted in the previous section, the words used in these two tweets, and retained in our filtering (to have sufficient number of occurrences), were listed. The word set shared by these two selected tweets consisted of the following 33 words: "to", "and", "it", "on", "you", "me", "not", "have", "up", "too", "from", "good", "well", "think", "ve", "been", "may", "much", "twitter", "fun", "brumplum", "enough", "everyone", "give", "obviously", "aggression", "around", "bye", "convinced", "henceforward", "pity", "retire", "unkindness".

Then we seek out all other tweets that use at least one of these words. That resulted in 225 out of the total of 302 tweets being retained.

Figure 2.1 positions our two critical tweets in a best planar projection of the tweets and associated words. In Figure 2.1, the contribution to the inertia of factor 1 by the "aggression" tweet is the greatest among all tweets, and the contribution to the inertia of factor 2 by the "I resign" tweet is the greatest among all tweets. While useful for finding dominant themes (expressed by the words used in the tweets), and perhaps also for the trajectory of these themes, we can use all of the latent semantic representation of this Twitter data set by clustering the tweets, based on their (Euclidean metric) factor projections.

Figure 2.2 displays this expression of the Twitter narrative here. Our crucial tweets are located at the end of a fairly compact structure. This points to how our two crucial tweets can be considered as bringing a sub-narrative to a conclusion. Our interest is therefore raised in finding sub-narratives in the Twitter flow. These sub-narratives are sought here as chronologically successive tweets, that is, a segment in the chronological flow of tweets.

2.1.3   Two Critical Tweets in Terms of Twitter Sub-narratives

To investigate our two critical tweets, such as the immediate or other past antecedents, and the repercussions or subsequent evolution of the Twitter narrative, we will now determine sub-narratives in the overall Twitter narrative. This we will do through segmentation of the

Principal factor projection of tweets in 33-word space

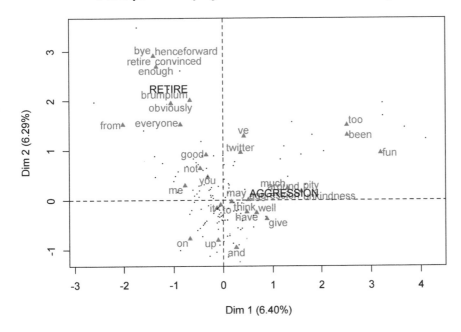

FIGURE 2.1: Factors 1 and 2, the best two-dimensional projection of the data cloud of 302 tweets, and of the 33-word set. The dots are at the locations of the tweets (identifiers are not shown, to avoid overcrowding). Just two tweets, the crucial two, are shown with the "retire" and "aggression" labels.

flow of tweets. So a sub-narrative is defined as a segment of this flow of tweets. That is, the sub-narrative consists of groups (or clusters) of successive tweets that are semantically homogeneous. Semantic homogeneity is defined through a statistical significance approach.

Sub-narratives, Twitter Data Used, Hierarchical Structure of the Overall Twitter Narrative

On the full set of tweets and the words used in these tweets, a threshold of 5 tweets was required for each word, and the total number of occurrences of words needed to be at least 5. This lowered our word set, initially 1787, to 143. Then we removed stopwords, and partial words, in a list that we made: "the", "to", "and", "of", "in", "it", "is", "for", "that", "on", "at", "be", "this", "what", "an", "if", "ve", "don", "ly", "th", "tr", "ll". That led to 121 words retained. Since some tweets were not lacking in words, there remained 280 non-empty tweets (from the initial set of 302 tweets). Our two critical tweets (the "I retire" and the "aggression" ones) were among the retained tweet set.

Following a full-dimensionality correspondence analysis, an agglomerative hierarchical clustering was applied on the factor space coordinates (hence endowed with the Euclidean metric). The chronological sequence of tweets was hierarchically clustered (see [188, 19] for this hierarchical clustering approach). With the set of 280 tweets, crossed by the 121 words, Figure 2.3 shows the chronological hierarchical clustering. Our two critical tweets are in their chronological sequence in the 280-tweet sequence (at the 211th and 212th tweet positions in this sequence).

Sequence-constrained hierarchical clustering of 225 tweets

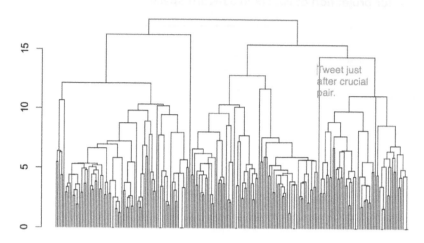

FIGURE 2.2: Hierarchical clustering, using the complete link agglomerative criterion (good for compact clusters) on the full-dimensionality Euclidean factor coordinates. Just 33 words are used. The tweet is annotated that is immediately following the two crucial ones that we are focused on: the "I retire" tweet and the "aggression" tweet. It is still a singleton until merged relatively late into the sequence of agglomerations.

A note now follows on why we did not use hashtag words (themes referred to), or @-symbol-prefaced words (other tweeters by Twitter name). The hashtag was not used all that often, the usages being: #media140, #thearchers, #frys, #FryS, #140conf, #grandslamdarts, #pdc (previous two generally together), #webwar, #threestrikes (previous two together always), #svuk. The total number of @-symbol names was 86. This was insufficient to base our entire analysis on Twitter names, even if hashtag themes were added.

To exploit the visualization of the Twitter narrative that is expressed in Figure 2.3, we will summarize this visualization by determining a segmentation of the flow of tweets. That is equivalently expressed as determining a partition of tweets from the dendrogram. Furthermore, as described in the next subsection, we look for internal nodes of the dendrogram that are statistically significant (using the approach that will now be described).

Sub-narratives of the Overall Twitter Narrative through Segmenting the Twitter Flow

In line with [19], we made these agglomerations subject to a permutation test to authorize (or not) each agglomeration that is deemed to be significant. In the description that now follows for determining significant segments of tweets, we follow very closely [19]. Statistical significance means that the agglomerands validly form a single segment.

All the distances between pairs of objects of the two adjacent groups that are candidates for agglomeration are computed. These distances are divided into two groups: 50% of the distances with the highest values are coded "1", and 50% with the lowest values are coded "0". The count of high distances in the between-group matrix (shared area between objects of one group and the other) is computed and denoted by h. The count of high distances between permuted groups is computed.

The number of permutations producing a result equal to or greater than h, divided by

Hierarchical clustering of 280 tweets in sequence

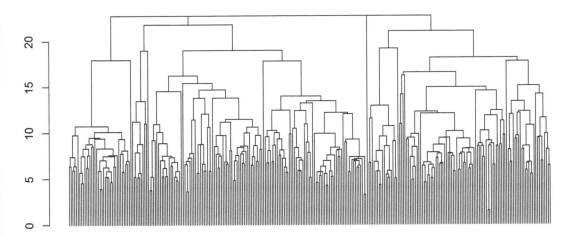

FIGURE 2.3: Hierarchical clustering, using the complete link agglomerative criterion (good for compact clusters) on the full-dimensionality Euclidean factor coordinates. The tweets are characterized by presence of any of the 121 words used. The 280 tweets, in chronological sequence, are associated with the terminal nodes (arranged horizontally at the bottom of the dendrogram or hierarchical tree). We will look for an understanding of semantic content, and the evolution of this, leading up to our two crucial tweets, and the further evolution of the tweet flow.

the number of permutations that are performed, gives an estimate of the probability p of observing the data under the following scenario that defines the null hypothesis: the objects in the two groups are drawn from the same statistical population and, consequently, it is only an artefact of the agglomerative clustering algorithm that they temporarily form two groups.

Thus, the proportion of permutations more extreme than the observed data is obtained. This proportion is termed the p-value of the permutation. The probability p is compared with a pre-established significance level α. If $p > \alpha$, the null hypothesis is accepted, the two groups can be said to be "homogeneous", and the fusion of the two groups is carried out. If $p \leq \alpha$, the null hypothesis is rejected, the two groups can be said to be "heterogeneous", and fusion of the groups is prevented. Changing the value of α changes the resolution of the partition obtained, which is what is obtained when the sequence of agglomerations is not allowed to go to its culmination point (of just one cluster containing all entities being clustered).

An α significance level of 0.15 was set (giving an intermediate number of segments between not too large, if α were set to a greater value, and a small number of segments, if the significance level were more demanding, i.e. smaller in value), and significance assessed through 5000 permutations (found to be very stable relative to using a few hundred permutations). The number of segments found was 40.

A factor space mapping of these 40 segments was determined, in their 121-word space. Four of these segments (the 6th, 18th, 36th and 39th) had just one tweet. Since they would therefore quite possibly perturb the correspondence analysis, in being exceptional in this way, we took these particular tweets as supplementary tweets. This means that the correspondence analysis factor space (i.e. the latent semantic space endowed with the

36 active clusters (40 in all) in factors 1, 2 plane

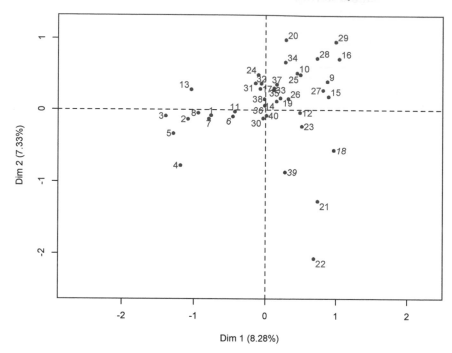

FIGURE 2.4: The centres of gravity of 40 segment groups of the Twitter flow are projected in the principal factor plane. See text for details related to 36 of these tweet segments being used for this analysis, and then four being projected into the factor space as supplementary tweets. These are 18 and 39 in the lower right quadrant, 36 very close to the origin, and 6 at the top of the lower left quadrant.

Euclidean metric) was determined using the active set of 40 less these four tweets, and then the four supplementary tweets were projected into the factor space.

The mapping is shown in Figure 2.4. It is noticeable that segment group 30, which contains our critical tweets towards the end of it, is very close to the origin, which is the average tweet here. The average tweet can be taken as the most innocuous. Therefore the factor plane of factors 1 and 2 is not useful for saying anything further about segment group 30, beyond that fact that it is entirely unremarkable.

The percentage contributions of segment group 30 to the factors 1, 2, 3, 4, 5 are, respectively 0.04, 1.71, 9.94, 1.62, 0.27. We will look at factors 2 and 3 because they are determined far more (than the other factors here) by segment group 30.

Figure 2.5 displays the words that are of greater contribution to the mean inertia of factors 2 and 3. We note that Figure 2.6 displays the important tweet segments in the plane of factors 2 and 3, that is, the tweet segments with contribution to the inertias of these factors that are greater than the mean. The chronological trajectory linking these important tweet segments is also shown.

Early tweet segments are positive on factor 3. Then there is a phase (with important tweet segments 21, 22, 25) that are fairly neutral on factor 3, but range first negatively/leftward on factor 2, and then positively/rightward. Then comes a phase (through tweet segment 27) of strong factor 3 positivity. Remember that the positive and negative

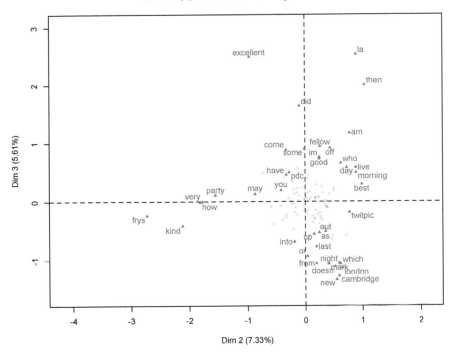

FIGURE 2.5: Important words, with contribution to the inertia of the cloud of all words in this factor plane, of factors 2 and 3. At top right `la` is LA, Los Angeles.

orientation of factor axes is only relative and contains no judgemental character whatsoever. With tweet segment 28 there is a move, reinforced by tweet segment 30 containing our crucial two tweets, back towards the other extremity of factor 3. Further tweet segments then play out their roles on the negative/southside factor 3 half-axis.

Segment groups positive on factor 3 were: 27, appearance in LA; 28, relating to a guest appearance on the show *Good Day LA*; and 20, recording of *Doctor Who*.

Segment groups negative on factor 3 were: 34, Cambridge (England) and (London Street) Norwich; 37, London (England); 32, concerning computer-related purchases and issues, and a London event; and 33, relating to Royal Geographic Society and other events.

So our tweet segment of interest, segment group 30, is between LA and the London area. In segment group 30, there is the alternation with Twitter user `@brumplum`, and also a mention of having arrived in LA. We note therefore these geographic linkages in Twitter vicinity of the crucial tweets relating to "aggression" and "I retire". Furthermore, we note the transition back to the London area, where events that Stephen Fry was involved in were based.

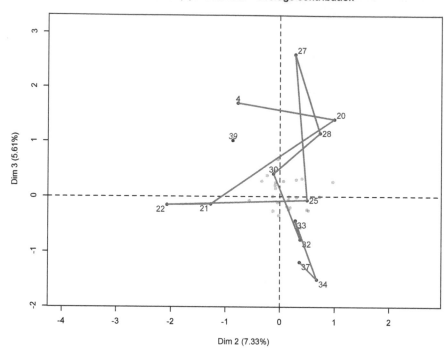

FIGURE 2.6: The plane of factors 2 and 3 with the important tweet segments. These tweet segments are important due to greater than mean contribution to the inertia of the cloud of tweet segments.

2.2   Analysis and Synthesis, Episodization and Narrativization

Examples at issue here, in the geometric and topological perspectives, through mapping, have used film and TV scripts in the converging world of cinema, television, games, online or virtual societies, with applications in entertainment, education and many other areas. For movies, see [188]; for *CSI: Las Vegas*, the crime scene investigation TV series, see [189]; and for novels, see [187]. In this context, our primary focus has been on the tracking of narrative. In this sense, analysis is the primary aim.

Next in this chapter, we describe how we can set about laying a basis for synthesis. That can be the generation of narratives, in various contexts, including novels and film scripts [190, 209]. The chapter concludes with a short account of various books that were written using what is described here as a support environment. Further scope and potential will be described a little too.

We explore (i) features or attributes of narrative that we can determine and measure, based on semantic analysis, using the movie *Casablanca*; (ii) segmentation – "episodization" is the term used in Aristotle's *Poetics* [7] (335 BC) – and its use, illustrated by cases from the *CSI: Las Vegas* television series; and (iii) narrative synthesis, including both a new technical approach to this, and our experience with a collaborative narrative creation "sandpit" environment. We can consider both analysis (i.e. breaking up of text units) and synthesis (i.e. assembling texts into a larger text unit).

In Aristotle the component parts of a play, for example, are considered. Under the heading "Outlines and episodization', Aristotle says: "Stories ... should first be set out in universal terms ... on that basis, one should then turn the story into episodes and elaborate it." He continues: "reasoning is the speech which the agents use to argue a case or put forward an opinion". It is interesting how the decomposition into episodes is related to agents who provide sense and meaning to the component parts. This is a theme which we will return to below. It may be noted (cf. White [244]) that Aristotle's perspectives on story and narrative retain importance to this day.

In narrativization we seek to build a story-line from an arbitrarily large number of texts. We can achieve a fixed target story-line size or length. There is traceability of all component parts of the story-line (which, we may note, is potentially important for propagation and preservation of digital rights). By integrating story transitions with heuristics for structuring story we allow additional story sub-plot embedding, and the placement of other relevant information. Furthermore, by appropriate mapping of story transitions, we can allow interactive reading and other adaptive usage frameworks.

2.3 Storytelling as Narrative Synthesis and Generation

Papadimitriou [201] succinctly provides a wide range of examples of where storytelling is or should be central in teaching. He ends with an example from the celebrated novelist and philosopher Umberto Eco (1932 2016), termed "salgarism" by the latter, the "telltale symptoms" of which are "incongruity and discontinuity between story and embedded information". In our work we seek to find algorithmically such patterns or trends in the data. We do this, it must be noted, not just on a word level, or a sentence level, or a paragraph level, but rather in close association with some level or levels of information resolution. Regarding the latter (level of information resolution), it will be observed in this chapter that it is based on clusters of words, which may or may not be taking order or sequence of word placement into account, and which are typically based on a hierarchical clustering so that resolution scale in this context can be associated with level in such a hierarchy.

When one considers multiple platforms, encompassing media, format, and form and extent of interactivity, Murray [161] raises the question of what platform and/or delivery mechanism is superior. In answering this, she points to our often very simplistic understanding of story. Instead, to explore commonality across delivery platform, it is necessary, she notes, to capture meaning (i.e. semantics).

We will next look at recent work in the area of games-related authoring software, which is also a prime objective of what is described here.

More traditional forms of authoring software have been largely focused on identifying and using interactions between characters, events, authors and the like [147]. More recent approaches to narrative synthesis include Louchar et al. [147] for whom there is a "narrative paradox" in the imputed distance between plot and interaction, in so far as plot implies abstracting away from, and in a sense pulling against, the interactive environment. This leads to a focus on the process of creating a plot narrative, rather than the structure of the story. As Louchart et al. [147] note: "In EN [emergent narrative] we try to remove the need to 'think in terms of plot', because the notion of plot ... has a problematic tension with the role of the interactor." In this book, instead, we focus precisely on the structure of narrative, and show how readily we can reconcile this approach with an interactive environment (see Section 2.7 below). Kriegel and Aylett [127] refer to bottom-up collaborative authoring based on such an approach as remaining "incoherent and chaotic", which is not the case with our approach.

Aylett's emergent narrative (EN) approach is developed too by Swartjes and Theune [227], based on branching and causal networks in the context of an "overall space of possible stories". Authors can modify or accept the choice of story elements and prolongations proposed by a story generator (referred to as, respectively, debugging or co-creation). In concluding, Swartjes and Theune [227] question whether their own approach is capable of scaling. Further work in the symbolic artificial intelligence tradition can be seen in [228]. In the approach here, on the other hand, our main possible limitation on scaling is due to (computationally cubic-order) eigenreduction, which, because this is carried out on very sparse matrices, is unlikely to be a limitation for computational work on large sparse matrices (see [193]). We note, though, that our current and ongoing work is often with many millions of text segments, such as from Twitter.

In this section, rather than looking at narrative applied specifically to games, we are looking to use algorithmic techniques, which are also very relevant for games, to find and depict the narrative thread in screenplay, by defining the times when a screenplay becomes most like and most different from itself. We use screenplay as our testbed or environment.

A so-called "linear" story consists of conflicts or moments of decision, and it is our contention that using textual narrative to identify the deep structure of a screenplay will be extremely helpful in thinking about the development of serious games. A screenplay in fact is written to maximize the number and impact of those nodal narrative points which are, or will be, experienced by the audience in real time.

A problem of the tree structure model prevalent in game narratology (see the discussion above; see also [227, 228]), is that it does not recognize that the experience of any audience member will be to experience their individual journey as a single pathway regardless of how many possible pathways they might have taken.

Indeed, the notion of foregrounding story over plot is not just a function of the interactive environment. Stephen King, the novelist, whose work has inspired more successful original films than perhaps any other writer working today, has written that "plotting and the spontaneity of real creation aren't compatible" [122, p. 164]. He calls plot "a dullard's first choice. The story which results from it is apt to feel artificial and labored". So instead of starting with plot, King puts his characters in a situation and likes to "watch them try to work themselves free".

The dramatic narrative, including a novel or film script, is devised to model those moments of choice, and to involve the audience as an active collaborator in imagining the story, which involves all the future outcomes desired and feared by the audience and not just the narrative choices which actually occur in the script or film. See Ganz [87] for further discussion of this.

By developing a toolset or processing environment that allows narratives to be represented numerically and symbolically, we are finding the possible patterns of narrative desired or created by the audience. Some of this work has been accomplished up to now by measuring reaction, but it has not been done by analysing the text itself and using these analysis outcomes to create heuristics for narrative in a form that can be recognized by a machine. These heuristics could also well create parameters for narrative in serious games and, for example, be subsequently used to close down narrative options at certain stages of a game.

2.4 Machine Learning and Data Mining in Film Script Analysis

Under discussion in this section are the complementary roles of machine learning as well as data mining. We have initially chosen to look at film script as a particular kind of text, both as a text in itself which is broken up into discrete units, with considerable amounts of easily recognizable metadata attached to each scene, but also as a text which is intended to achieve itself in a different visual form which will be experienced chronologically and in real time.

A film script, expressing a story, is the starting point for any possible production for cinema or TV. See the short extract in Figure 2.7. TV episodes in the same series may each be developed by different scriptwriters, and later by different producers and directors. The aim of any TV screenplay is to provide a unique but repeatable experience in which each episode shares certain structural and narrative traits with other episodes from the same series, despite the fact they may have been originated or realized by different people or teams. There is a productive tension between the separate needs for uniqueness, such that each episode seems fresh and surprising, but also belongs to its genre. For example, an episode of the TV sitcom *Friends* needs to feel *Friends*-like, to offer the specific kinds of pleasure the audience associates with the series. We propose that these distinctive qualities of any individual script and the distinctive qualities of any genre are open to analysis through a pattern and trend analysis toolset that finds distinctive ways of representing the essential structural qualities of any script and the series to which it belongs, and thus enables the writer, the script developer or producer to have a deeper understanding of the script and have objective criteria for the creative decisions they take. Moreover, as the scripts are migrated to digital formats the tools offer many possibilities for prototyping from the information gathered.

By analysis of multiple screenplays, TV episodes and genres, the technology will allow the possibility of creating distinctive analytical patterns for the structure of genres, series, or episodes in the same way that comparative authorship can be assessed for individual writers. Large numbers of film scripts, for all genres, are available and openly accessible online (e.g. IMSDb, The Internet Movie Script Database, www.imsdb.com; or www.kilohoku.com).

A film script is a semi-structured textual document in that it is subdivided into scenes and sometimes other structural units. Both character dialogue and descriptive and background metadata are provided in a film script. The metadata are partially formalized since there are some invariants in any film script including set of characters; and essential metadata related to location (internal, external; particular or general location name), characters, and time of day. Accompanying the dialogue there is often description of action in free text. While offering just one data modality (text), there is close linkage to other modalities (visual, speech and sometimes music).

A substantial part of this book is focused on data mining, understood as unsupervised learning or knowledge discovery.

It is interesting to note that another important line of inquiry has been pursued in the supervised or machine learning direction. The feasibility of using statistical learning methods in order to map a feature space characterization of film scripts (using two dozen characteristics) onto box office profitability was pursued by [73]. The importance of such machine learning of what constitutes a good-quality and/or potentially profitable film script has been the basis for (successful) commercial initiatives. The business side of the movie business is elaborated on in some depth in [74]. The machine learning approach attempts to map film script inputs to target outputs (in practice, there is considerable manual intervention involved in data preprocessing). An important part of our data mining approach

[INT. CSI - EVIDENCE ROOM – NIGHT]
(WARRICK opens the evidence package and takes out the shoe.)
(He sits down and examines the shoe. After several dissolves, WARRICK opens the lip of the shoe and looks inside. He finds something.)
WARRICK BROWN: Well, I'll be damned.
(He tips the shoe over and a piece of toe nail falls out onto the table. He picks it up.)
WARRICK BROWN: Tripped over a rattle, my ass.

FIGURE 2.7: Example of a scene from a script. This is a short scene, scene 25, from the *CSI: Las Vegas* TV series. This is the very first, 1X01, pilot, originally broadcast on CBS on 6 October 2000. Written by A.E. Zuiker, directed by D. Cannon. Script available in full from TWIZ TV (Free TV Scripts & Movie Screenplays Archives), twiztv.com

attempts to find faint patterns and structure in the film script data that can be exploited for a range of decision-making, including such learning of mapping onto box-office success. Thus it is complementary to, and a base for, such commercial objectives.

2.5   Style Analytics: Statistical Significance of Style Features

For style analytics, our primary interest is in the style features that can be statistically analysed. These style features include climax, rhythm and tempo.

In [188], we carried out a detailed study of features proposed and discussed in [153]. Analysis of style and structure was carried out, using 999 randomizations of scenes, thereby furnishing what is needed for testing the statistical hypothesis of significance at the 0.001 level, from the sequence of 77 scenes in the movie *Casablanca*. In all, nine attributes were used. Through the randomizations of scene sequence we have a Monte Carlo approach to test statistical significance of the given script's patterns and structures as opposed to randomized alternatives.

Recall our discussion in Section 1.2.9 of the so-call "mid-act climax", scene 43, of *Casablanca*, which McKee [153] divides into 11 sub-scenes or "beats". The length of the beat can show a lead-up to a climax in the scene. We see this very well in the beats of scene 43: the final five beats have lengths (in terms of presence of the words we use) of 50, 44, 38, 30, and then in the climax beat, 46. Earlier beats vary in length, with successive word counts of 51, 23, 99, 39, 30, 17.

Given the projections of scenes, or sub-scenes or whatever document units are used, we can easily investigate such indicators of style and structure as the following.

1. Attributes 1 and 2: the relative movement, given by the mean squared distance from one scene to the next. We take the mean and the variance of these relative movements. Attributes 1 and 2 are based on the (full-dimensionality) factor space embedding of the scenes.

2. Attributes 3 and 4: the changes in direction, given by the squared difference in correlation from one scene to the next. We take the mean and variance of these changes in direction. Attributes 3 and 4 are based on the (full-dimensionality) correlations with factors.

3. Attribute 5 is mean absolute tempo. Tempo is given by difference in beat length

from one scene to the next. Attribute 6 is the mean of the ups and downs of tempo.

4. Attributes 7 and 8 are, respectively, the mean and variance of rhythm given by the sums of squared deviations from one scene length to the next.

5. Finally, attribute 9 is the mean of the rhythm taking up or down into account.

For scene 43 of *Casablanca* we found the following particularly significant. We tested the given scene, with its 11 beats, against 999 uniformly randomized sequences of 11 beats. If we so wish, this provides a Monte Carlo significance test of a null hypothesis up to the 0.001 level.

- In repeated runs, each of 999 randomizations, we find scene 43 to be particularly significant (in 95% of cases) in terms of attribute 2: variability of movement from one beat to the next is smaller than randomized alternatives. This may be explained by the successive beats relating to the coming together, or drawing apart, of Ilsa and Rick, as we have already noted.

- In 84% of cases, scene 43 has greater tempo (attribute 5) than randomized alternatives. This attribute is related to absolute tempo, so we do not consider whether decreasing or increasing.

- In 83% of cases, the mean rhythm (attribute 7) is higher than randomized alternatives.

We found the following, in summary, in this Monte Carlo "baselining" analysis. The entire plot of *Casablanca* is well characterized by the variability of movement from one scene to the next (attribute 2). Variability of movement from one beat to the next is smaller than randomized alternatives in 82% of cases. Similarity of orientation from one scene to the next (attribute 3) is very tight (i.e. smaller than randomized alternatives); we found this to hold in 95% of cases. The variability of orientations (attribute 4) was also tighter, in 82% of cases. Attribute 6, the mean of ups and downs of tempos, is also revealing: in 96% of cases, it was smaller in the real *Casablanca* than in the randomized alternatives. This points to the "balance" of up and down movement in pace.

2.6 Typicality and Atypicality for Narrative Summarization and Transcoding

We consider now the use of typicality and of atypicality with the general objectives either of summarization of narrative content, or transcoding. By transcoding, it can be assumed that very often narrative has multiple expressions, or manifestations. In Section 1.2.2 note was made of the 360-degree media world, including quite possibly both multimedia and social media. In Chapter 8 there will be a link-up with new and important developments that integrate quantitative and qualitative analytics in psychoanalysis.

Taking Karl Marx's account of commodity fetishism [151], we subdivided it into 21 consecutive paragraphs. Word counts varied from 512 words in paragraph 6 to just 25 words in the one-sentence paragraph 5.

Figure 2.8, using all the data, gives rise to the following. Most paragraphs are bunched near the origin, that is, the average paragraph profile. (One speaks of "profiles" in correspondence analysis because paragraphs – rows – are normalized in the analysis to even out the different numbers of terms per paragraph. Similarly, term – column – profiles arise

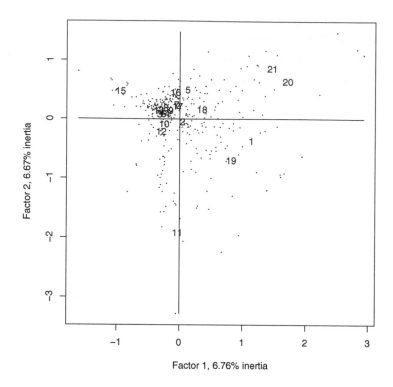

FIGURE 2.8: Recounting of commodity fetishism. Projections on this planar visualization of paragraphs 1–21 are noted. The 974 terms that characterize these paragraphs are displayed as dots. Percentage inertia quantifies the relative information of the factor or axis.

through the same evening out of the different numbers of paragraphs characterized by each term.)

So the paragraphs that seem to really matter are 15, 21, 20, 1, 19 and 11. We can quantify what we mean by "really matter" through the mathematically defined correlations and contributions to the factors.

As a consequence of these findings in Figure 2.8, we will proceed with just the restricted set of six paragraphs as constituting the "backbone" of Karl Marx's recounting. Figure 2.9, using six paragraphs instead of the given 21, and hence with 482 associated terms rather than 974 associated terms, is not unlike Figure 2.8. It is important to note that the factors of correspondence analysis are not fixed in their orientations. Hence, it is equally acceptable to have reflections in the axes, just so long as there is consistency. So in both figures, paragraphs 20 and 21 are closely located. These two paragraphs are counterposed to paragraph 11. In both figures, paragraphs 20 and 21, and then 11, and finally 15, constitute approximately three vertices of a triangle. This is a homology between these two figures. What does change in location in proceeding to Figure 2.9 from Figure 2.8 is paragraph 19; paragraph 19 is repositioned from being between 21, 20 and 11, to being between 21, 20 and 15. Paragraph 1 has come a little closer to the origin (and hence the average paragraph).

If one wished to have the most salient paragraphs to summarize the overall text, we therefore find this job to be done quite well by using paragraphs 1, 11, 15, 19, 20 and 21.

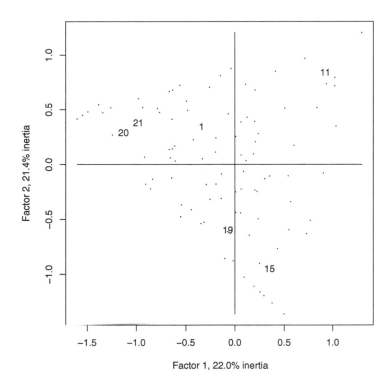

FIGURE 2.9: Compared to Figure 2.8, a smaller set of paragraphs is used here: paragraphs 1, 11, 15, 19, 20, 21. There are 482 terms in these paragraphs, displayed as dots.

The reader of the text on commodity fetishism who is in a hurry should concentrate on these paragraphs! In brief, the contents of the paragraphs are as follows.

1. Paragraph 1: commodity's attributes
2. Paragraph 11: Robinson Crusoe story
3. Paragraph 15: religion and worship
4. Paragraph 19: on the provenance of the monetary system
5. Paragraphs 20, 21: use value and/or exchange value.

Marx's writing can be very visual, with plenty of illustrations. For instance, in paragraph 1, there is reference to tables made from wood. In this writing, metaphor and metonymy are very prominent.

What we have exemplified in this case study is semantically based narrative summarization. Apart from summarization, this analysis approach can be used also for selecting core visual metaphors or other features for multiplatform transcoding.

As regards visual metaphors needed in taking the script into the film set, or performance environment, text in time is designed to create visual metaphors in the teller and the audience, as Ganz [87] has explored in writing about the links between oral narrative and the cinema. Ganz [88], in a chapter in a book on storytelling in world cinema, considered how it is the dynamic invocation of images that is an essential element of oral narrative.

These ideas are also discussed in the work of Elizabeth Minchin, cited in [190]: "The story-teller's language serves as a prompt and a guide to stimulate us to perform the exercise of visualization and to ensure that the picture which we build up is appropriate."

2.7   Integration and Assembling of Narrative

Narrativization is how we term the finding and tracking of the narrative. An approach was developed for a support environment for distributed literary creation. In the "Too Many Cooks Project", to start with, students in a class on creative writing, in the discipline of English language, collectively wrote a novel in a week. This was printed and published in 2009 as a 231-page Kindle e-book, *The Delivery*, under the the pseudonym Tim Cooks. Many other such e-books, collectively authored by schoolchildren in schools in the London (UK) area, were published with collective author name Tim Cooks, and under the guidance of Joe Reddington. A YouTube video on this is available at `www.narrativization.com`. This video overviews the "workflow of the novel writing process", by going "deep into the style structure" and drawing out relevant information in regard to "narrative structure and changes in style".

The support environment here was for collaborative, distributed creating of narrative. This includes pinpointing anomalous sections; assessing homogeneity of style over successive iterations of the work; scenario experimentation and planning; condensing parts, or elaborating; and evaluating similarity of structure relative to best practice in the chosen genre.

All of the methodological work on narrativization is relevant for general information spaces, including all forms of human–machine interaction.

A narrative functions effectively because in simple terms it suggests a causal or emotional relationship between events. A story is an expression of causality or connection. Its aim is the transformation of a number of discrete elements (facts or views or other units of information) into part of a composite – the narrative. This process also involves rejecting those elements that are not germane to that narrative. Thus a narrative simultaneously binds together those elements that are germane to it and, through a process of editing and retelling, discloses those that are not part of that particular narrative. This process is discussed in Bruner [40].

We have been exploring how the methods can be used to structure user-generated content "narratively". This includes, firstly, the ordering or reordering of contributions. We want these contributions to best represent a narrative that can be read from the totality of contributions to any particular thread of activity (including, for example, comments on news stories or posts in online forums). Secondly, we seek to discard or render invisible contributions that do not take the narrative forward. Any set of contributions to a thread are to be arranged beyond chronological and thematic structuring to create narrative, using salience and interestingness detection. Finally, we look at those contributions that change the future contributions to a thread (just as the methods have been exploited to detect the turning points and act breaks in a screenplay).

In Theune et al. [231], an algorithmic, distributed narrative story engine is described. Modules are responsible for plot, narrative and presentation. However, plot and narrative are interlinked in an intricate way. So in our work we want this inherent relationship to be maintained. Implementationwise, we avail of an interactive support environment. We can essentially say that in our approach the semantics can be traced out, or demarcated, through the syntax elements, and reciprocally (not least through our use of "tool words") the semantics can be manifested above and beyond the syntax. We can note therefore how, for us, plot, narrative and presentation are all interwoven, and how we address this.

This very successful application shows how quantitative and qualitative aspects of narrative analysis and synthesis can be well integrated, with quite profound consequences for outputs that are achieved and accomplished.

Part II

Foundations of Analytics through the Geometry and Topology of Complex Systems

3

Symmetry in Data Mining and Analysis through Hierarchy

3.1 Analytics as the Discovery of Hierarchical Symmetries in Data

In this chapter, we have data analytics as the discovery of symmetries in data. This well addresses our contemporary big data needs, especially because symmetries can be at different resolution scales. That is to say, we may consider the overall and general issues either observed or confronting us. We may also consider the specific issues in detail. Resolution scale is nicely expressed as hierarchy. A chapter in an important book by 1978 Nobel Prize winner, Herbert Simon [219], has the appropriate chapter title "The architecture of complexity: hierarchic systems".

Hierarchy provides a unifying view of patterns, in the context of data mining and data analysis. We consider how hierarchy fully fulfils the role of determining data symmetries.

Symmetry plays a fundamental role in theoretical physics and in many other domains like art and design. Group theory is the way that mathematics views symmetries. Here we describe the various ways that hierarchy, and related data analysis and data handling, express symmetries in data. This provides a good background for later discussion, and in particular in Part IV.

Data analysis and data mining are concerned with unsupervised pattern finding and structure determination in data sets. The data sets themselves are explicitly linked as a form of representation to an observational or otherwise empirical domain of interest. "Structure" has long been understood as symmetry which can take many forms with respect to any transformation, including point, translational, rotational, and many others. Symmetries directly point to invariants, which pinpoint intrinsic properties of the data and of the background empirical domain of interest. As our data models change, so too do our perspectives on analysing data. The structures in data surveyed here are based on hierarchy, represented as p-adic numbers or an ultrametric topology.

3.2 Introduction to Hierarchical Clustering, p-Adic and m-Adic Numbers

Herbert A. Simon, Nobel Laureate in Economics, originator of "bounded rationality" and of "satisficing", believed in hierarchy at the basis of the human and social sciences, as the following quotation shows: "my central theme is that complexity frequently takes the form of hierarchy and that hierarchic systems have some common properties independent of their specific content. Hierarchy, I shall argue, is one of the central structural schemes that the architect of complexity uses" [219, p. 184].

Partitioning a set of observations [225, 226, 157] leads to some very simple symmetries.

This is one approach to clustering and data mining. But such approaches, often based on optimization, are really not of direct interest to us here. Instead we will pursue the theme pointed to by Simon, namely that the notion of hierarchy is fundamental for interpreting data and the complex reality which the data expresses. Our work is very different too from the marvellous view of the development of mathematical group theory – but viewed in its own right as a complex, evolving system – presented by Foote [78].

3.2.1 Structure in Observed or Measured Data

Weyl [243] makes the case for the fundamental importance of symmetry in science, engineering, architecture, art and other areas. As a "guiding principle", "[w]henever you have to do with a structure-endowed entity ... try to determine its group of automorphisms, the group of those elementwise transformations which leave all structural relations undisturbed. You can expect to gain a deep insight in the constitution of [the structure-endowed entity] in this way. After that you may start to investigate symmetric configurations of elements, i.e. configurations which are invariant under a certain subgroup of the group of all automorphisms" [243, p. 144].

"Symmetry is a vast subject, significant in art and nature.", Weyl states (p. 145), and there is no better example of the "mathematical intellect" at work. "Although the mathematics of group theory and the physics of symmetries were not fully developed simultaneously – as in the case of calculus and mechanics by Newton – the intimate relationship between the two was fully realized and clearly formulated by Wigner and Weyl, among others, before 1930" [234, p. 1]. Powerful impetus was given to this (mathematical) group view of study and exploration of symmetry in art and nature by Felix Klein's 1872 Erlangen Programme [123] which proposed that geometry was at heart group theory: geometry is the study of groups of transformations, and their invariants. Klein's Erlangen Programme is at the crossroads of mathematics and physics. The purpose of this chapter includes the following: to locate symmetry and group theory at the crossroads of data mining and data analytics too.

A short listing of what comes later in this chapter is as follows. In Section 3.3, we describe ultrametric topology as an expression of hierarchy. In Section 3.4.2, p-adic encoding, providing a number-theory vantage point on ultrametric topology, gives rise to additional symmetries and ways to capture invariants in data. Section 3.6 deals with symmetries that are part and parcel of a tree, representing a partial order on data, or equally a set of subsets of the data, some of which are embedded. In Section 3.7 permutations are at issue, including permutations that have the property of representing hierarchy. Section 3.8 deals with newer results relating to the remarkable symmetries of massive, and especially high-dimensional, data sets.

3.2.2 Brief Look Again at Hierarchical Clustering

For the reader new to analysis of data a very short introduction is now provided on hierarchical clustering. Along with other families of algorithms, the objective is automatic classification, for the purposes of data mining or knowledge discovery. Classification, after all, is fundamental in human thinking and machine-based decision-making. But we draw attention to the fact that our objective is *unsupervised* as opposed to *supervised* classification, also known as discriminant analysis or (in a general way) machine learning. So here we are *not* concerned with generalizing the decision-making capability of training data, nor are we concerned with fitting statistical models to data so that these models can play a role in generalizing and predicting. Instead we are concerned with having "data speak for themselves". That this unsupervised objective of classifying data (observations, objects, events, phenomena, etc.) is a huge task in our society is unquestionably true. One may think of situations when precedents are very limited, for instance.

Among families of clustering, or unsupervised classification, algorithms, we can distinguish the following: (i) array permuting and other visualization approaches; (ii) partitioning to form (discrete or overlapping) clusters through optimization, including graph-based approaches; and – of interest to us in this chapter – (iii) embedded clusters interrelated in a tree-based way.

For the last mentioned family of algorithm, agglomerative building of the hierarchy from consideration of object pairwise distances has been the most common approach adopted. As comprehensive background texts, see [156, 99, 247, 100].

3.2.3 Brief Introduction to p-Adic Numbers

The real number system, and a p-adic number system for given prime, p, are potentially equally useful alternatives. p-adic numbers were introduced by Kurt Hensel in 1898.

The predominance of real numbers is based on starting with the natural numbers. Our measurements are based on, firstly, natural numbers, and then, from fractions, rational numbers, which become generalized to real numbers. Interesting views of how we carry out observational science can be found in Volovich [240, 239] (see also Freund [83]). Here, we assume that we use rationals to make measurements. But they will be approximate, in general. It is better therefore to allow for observables being 'continuous, that is, endow them with a topology. Therefore we need a completion of the field $\mathbb{Q}$ of rationals. To complete the field $\mathbb{Q}$ of rationals we need Cauchy sequences, and this requires a norm on $\mathbb{Q}$ (because the Cauchy sequence must converge, and the norm is the tool used to show this). There is the Archimedean norm such that, for any $x, y \in \mathbb{Q}$, with $|x| < |y|$, there exists an integer N such that $|Nx| > |y|$. For convenience here, we write $|x|_\infty$ for this norm. So if this completion is Archimedean, then we have $\mathbb{R} = \mathbb{Q}_\infty$, the reals. That is fine if space is taken as commutative and Euclidean.

What of alternatives? Besides the $\mathbb{Q}_\infty$ norm, we have an infinity of norms, $|x|_p$, labelled by primes, p. We have labelling, via p, of the infinite set of non-Archimedean completions of $\mathbb{Q}$ to a field endowed with a topology.

In all cases we obtain locally compact completions, $\mathbb{Q}_p$, of $\mathbb{Q}$. They are the fields of p-adic numbers. All these $\mathbb{Q}_p$ are continua. Being locally compact, they have additive and multiplicative Haar measures. As such we can integrate over them, such as for the reals.

3.2.4 Brief Discussion of p-Adic and m-Adic Numbers

We will use p to denote a prime, and m to denote a non-zero positive integer. A p-adic number is such that any set of p integers which are in distinct residue classes modulo p may be used as p-adic digits. (See the remark at the end of Section 3.5.1, quoting from [90]. It makes the point that this opens up a range of alternative notation options in practice.) Recall that a ring does not allow division, while a field does. m-adic numbers form a ring; but p-adic numbers form a field. So *a priori*, 10-adic numbers form a ring. This provides us with a reason for preferring p-adic over m-adic numbers.

We can consider various p-adic expansions:

1. $\sum_{i=0}^{n} a_i p^i$, which defines positive integers. For a p-adic number, we require $a_i \in 0, 1, \ldots, p - 1$. (As an example, just consider the use of $p = 2$, binary numbers.)

2. $\sum_{i=-\infty}^{n} a_i p^i$ defines rationals.

3. $\sum_{i=k}^{\infty} a_i p^i$, where k is an integer, not necessarily positive, defines the field $\mathbb{Q}_p$ of p-adic numbers.

$\mathbb{Q}_p$, the field of p-adic numbers, is (as seen in these definitions) the field of p-adic expansions. Further description of p-adic number usage is given in Section 4.8.

3.3   Ultrametric Topology

In this section we mainly explore symmetries related to geometric shape, matrix structure and lattice structures. Ultrametric topology is topology on a hierarchy or rooted tree.

3.3.1   Ultrametric Space for Representing Hierarchy

Consider Figure 3.1, illustrating the ultrametric distance and its role in defining a hierarchy. An early, influential paper is Johnson [104], and an important survey is that of Rammal et al. [207]. Discussion of how a hierarchy expresses the semantics of change and distinction can be found in [177].

The ultrametric topology dates from Marc Krasner [126], and the ultrametric inequality from Hausdorff in 1934. From [212]: real and complex fields gave rise to the study of any field K with a complete valuation $|\cdot|$ comparable to the absolute value function. Such fields are to satisfy the strong triangle inequality $|x + y| \leq \max(|x|, |y|)$. When we have a valued field, defining a totally ordered Abelian (i.e. commutative) group, this is endowed with an ultrametric space through $|x - y| = d(x, y)$. Various terms are used interchangeably for such analysis, including p-adic, ultrametric, non-Archimedean, and isosceles. The natural geometric ordering of metric valuations is on the real line, whereas in the ultrametric case the natural ordering is a hierarchical tree.

3.3.2   Geometrical Properties of Ultrametric Spaces

We see from the following, based on [137, Chapter 0, Part IV], that an ultrametric space is quite different from a metric one. In an ultrametric space everything "lives" on a tree.

In an ultrametric space, all triangles are either isosceles with small base, or equilateral. We have here very clear symmetries of shape in an ultrametric topology. These symmetry "patterns" can be used to fingerprint data data sets and time series; see [169, 172] for many examples of this.

Some further properties that are studied in [137] are as follows. (i) Every point of a circle in an ultrametric space is a centre of the circle. (ii) In an ultrametric topology, every ball is both open and closed (termed *clopen*). (iii) An ultrametric space is zero-dimensional (see [46, 237]). It is clear that an ultrametric topology is very different from our intuitive, or Euclidean, notions. The most important point to keep in mind is that in an ultrametric space everything "lives" (informally expressing "exists") in a hierarchy expressed by a tree.

3.3.3   Ultrametric Matrices and Their Properties

For an $n \times n$ matrix of positive reals, symmetric with respect to the principal diagonal, to be a matrix of distances associated with an ultrametric distance on a set X, a sufficient and

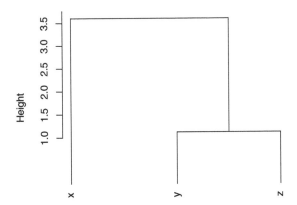

FIGURE 3.1: The strong triangle inequality defines an ultrametric: every triplet of points satisfies the relationship $d(x, z) \leq \max\{d(x, y), d(y, z)\}$ for distance d. See, by reading off the hierarchy, how this is verified for all x, y, z: $d(x, z) = 3.5$, $d(x, y) = 3.5$, $d(y, z) = 1.0$. In addition, the symmetry and positive definiteness conditions hold for any pair of points.

necessary condition is that a permutation of rows and columns satisfies the following form of the matrix.

1. Above the diagonal term, equal to 0, the elements of the same row are non-decreasing.

2. For every index k, if

$$d(k, k+1) = d(k, k+2) = \cdots = d(k, k+\ell+1)$$

then

$$d(k+1, j) \leq d(k, j), \quad \text{for } k+1 < j \leq k+\ell+1,$$

and

$$d(k+1, j) = d(k, j), \quad \text{for } j > k+\ell+1.$$

Under these circumstances, $\ell \geq 0$ is the length of the section beginning, beyond the principal diagonal, the interval of columns of equal terms in row k.

To illustrate the ultrametric matrix format, consider the small data set shown in Table 3.1. A dendrogram produced from this is given in Figure 3.2. The ultrametric matrix that can be read off this dendrogram is shown in Table 3.2. A visualization of this matrix, illustrating the ultrametric matrix properties discussed above, is on the right in Figure 3.2. It uses, not the given order of iris flowers, but instead the order resulting from the row sums in the ultrametric matrix, and these are identical to the column sums. That expresses, precisely here, a consistent left/right order of agglomerations in the agglomerative hierarchical clustering. That is, it is associated with dendrogram having such a consistent left and right child nodes of each parent node in the tree (or dendrogram, or hierarchy). This ordering is, with identical values for 3, 4 and 5, 6: 3, 4, 7, 2, 1, 5, 6.

	Sepal.Length	Sepal.Width	Petal.Length	Petal.Width
iris1	5.1	3.5	1.4	0.2
iris2	4.9	3.0	1.4	0.2
iris3	4.7	3.2	1.3	0.2
iris4	4.6	3.1	1.5	0.2
iris5	5.0	3.6	1.4	0.2
iris6	5.4	3.9	1.7	0.4
iris7	4.6	3.4	1.4	0.3

TABLE 3.1: Input data: 7 iris flowers characterized by sepal and petal widths and lengths. From Fisher's iris data [76].

	iris1	iris2	iris3	iris4	iris5	iris6	iris7
iris1	0.00	1.27	1.27	1.27	0.14	0.71	1.27
iris2	1.27	0.00	0.44	0.44	1.27	1.27	0.44
iris3	1.27	0.44	0.00	0.24	1.27	1.27	0.32
iris4	1.27	0.44	0.24	0.00	1.27	1.27	0.32
iris5	0.14	1.27	1.27	1.27	0.00	0.71	1.27
iris6	0.71	1.27	1.27	1.27	0.71	0.00	1.27
iris7	1.27	0.44	0.32	0.32	1.27	1.27	0.00

TABLE 3.2: Ultrametric matrix derived from the dendrogram in Figure 3.2.

3.3.4 Clustering through Matrix Row and Column Permutation

Figure 3.2 (right) shows how an ultrametric distance allows a certain structure to be visible (quite possibly, in practice, subject to an appropriate row and column permutation), in a matrix defined from the set of all distances. For set X, then, this matrix expresses the distance mapping of the Cartesian product, $d : X \times X \longrightarrow \mathbb{R}^+$, where $\mathbb{R}^+$ denotes the non-negative reals. *A priori* the rows and columns of the function of the Cartesian product set X with itself could be in any order. The ultrametric matrix properties establish what is possible when the distance is an ultrametric one. Because the matrix (a two-way data object) involves one *mode* (due to set X being crossed with itself; as opposed to the two-mode case where an observation set is crossed by an attribute set), it is clear that both rows and columns can be permuted to yield the *same* order on X. A property of the form of the matrix is that small values are at or near the principal diagonal.

A generalization opens up for this sort of clustering by visualization scheme. Firstly, we can directly apply row and column permuting to two-mode data, that is, to the rows and columns of a matrix crossing indices I by attributes J, $a : I \times J \longrightarrow \mathbb{R}$. A matrix of values, $a(i, j)$, is furnished by the function a acting on the sets I and J. Here, each such term is real-valued. We can also generalize the principle of permuting such that small values are on or near the principal diagonal to instead allow similar values to be near one another, and thereby to facilitate visualization. An optimized way to do this was pursued in [152, 150]. Comprehensive surveys of clustering algorithms in this area, including objective functions, visualization schemes, optimization approaches, presence of constraints, and applications, can be found in [235, 148]; see also [59, 167].

For all these approaches, underpinning them are row and column permutations which can be expressed in terms of the permutation group, S_n, on n elements.

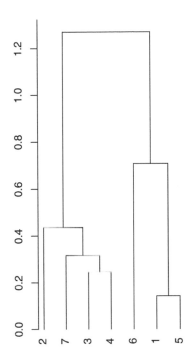
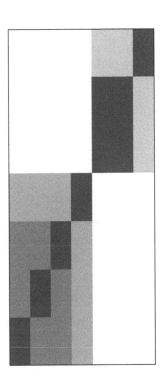

FIGURE 3.2: Hierarchical clustering of seven iris flowers using data from Table 3.1. No data normalization was used. The agglomerative clustering criterion was the minimum-variance or Ward one. On the right is a visualization of the ultrametric matrix of Table 3.2, where bright or white signifies highest value, and black lowest value. The ordering of rows (columns) is that of the order of row (column) sums in the ultrametric matrix. This ordering is: 3, 4, 7, 2, 1, 5, 6.

3.3.5 Other Data Symmetries

As examples of various other local symmetries worthy of consideration in data sets, consider subsets of data comprising clusters, and reciprocal nearest neighbour pairs.

Given an observation set, X, we define dissimilarities as the mapping $d : X \times X \longrightarrow \mathbb{R}^+$. A dissimilarity is a positive definite, symmetric measure (i.e. $d(x, y) \geq 0$, $d(x, y) = 0$ if $x = y$, $d(x, y) = d(y, x)$). If, in addition, the triangle inequality is satisfied (i.e. $d(x, y) \leq d(x, z) + d(z, y)$, for all $x, y, z \in X$) then the dissimilarity is a distance.

If X is endowed with a metric, then this metric is mapped onto an ultrametric. In practice, there is no need for X to be endowed with a metric. Instead a dissimilarity is satisfactory.

A hierarchy, H, is defined as a binary, rooted, node-ranked tree, also termed a dendrogram [25, 104, 137, 167]. A hierarchy defines a set of embedded subsets of a given set of objects X, indexed by the set I. That is to say, object i in the object set X is denoted x_i, and $i \in I$. These subsets are *totally ordered* by an index function ν, which is a stronger condition than the *partial order* required by the subset relation. The index function ν is represented by the ordinate in Figure 3.2 (the "height" or "level"). A bijection exists between a hierarchy and an ultrametric space.

Often in this chapter we will refer interchangeably to the object set, X, and the associated set of indices, I.

Usually a constructive approach is used to induce H on a set I. The most efficient algorithms are based on nearest neighbour chains, which by definition end in a pair of agglomerable reciprocal nearest neighbours. Various studies of computational complexity are available in [163, 164, 167, 168]. The author of this book provided the original implementations in the R software package and in Clustan.

3.4 Generalized Ultrametric and Formal Concept Analysis

In this section we consider an ultrametric defined on the power set or join semilattice. Comprehensive background on ordered sets and lattices can be found in [56]. A review of generalized distances and ultrametrics can be found in [215].

3.4.1 Link with Formal Concept Analysis

Typically hierarchical clustering is based on a distance (which can be relaxed often to a dissimilarity, not respecting the triangle inequality, and *mutatis mutandis* to a similarity), defined on all pairs of the object set: $d : X \times X \to \mathbb{R}^+$. That is, a distance is a positive real value. Usually we require that a distance cannot be zero-valued unless the objects are identical. That is the traditional approach.

A different form of ultrametrization is achieved from a dissimilarity defined on the power set of attributes characterizing the observations (objects, individuals, etc.) X. Here we have $d : X \times X \longrightarrow 2^J$, where J indexes the attribute (variables, characteristics, properties, etc.) set.

This gives rise to a different notion of distance, which maps pairs of objects onto elements of a join semilattice. The latter can represent all subsets of the attribute set, J. That is to say, it can represent the power set, commonly denoted 2^J, of J.

The reason for this notation is that the cardinality of the power set is $2^{|J|}$, where $|J|$ is the cardinality of J. Consider the case of $J = \{v_1, v_2, v_3\}$. Then all of the following, eight sets in all, are members of 2^J: $\emptyset, \{v_1\}, \{v_2\}, \{v_3\}, \{v_1, v_2\}, \{v_1, v_3\}, \{v_2, v_3\}, \{v_1, v_2, v_3\} = J$. So $|J| = 3$ and $2^{|J|} = 8$.

As an example, consider, say, $n = 5$ objects characterized by three boolean (presence/absence) attributes, shown in Figure 3.3 (top). Define dissimilarity between a pair of objects in this table as a *set* of three components, corresponding to the three attributes, such that if both components are 0, we have 1; if either component is 1 and the other 0, we have 1; and if both components are 1 we get 0. This is the simple matching coefficient [102]. We could use, for example, Euclidean distance for each of the values sought; but we prefer to treat 0 values in both components as signalling a 1 contribution. We then get $d(a, b) = 1, 1, 0$, which we will call d1,d2. Then $d(a, c) = 0, 1, 0$, which we will call d2. And so on. With the latter we create lattice nodes as shown in the middle part of Figure 3.3.

In formal concept analysis [56, 86], it is the lattice itself that is of primary interest. In [102] there is discussion of, and a range of examples on, the close relationship between the traditional hierarchical cluster analysis based on $d : I \times I \to \mathbb{R}^+$, and hierarchical cluster analysis "based on abstract posets" (a poset is a partially ordered set), based on

	v_1	v_2	v_3
a	1	0	1
b	0	1	1
c	1	0	1
e	1	0	0
f	0	0	1

```
Potential lattice vertices      Lattice vertices found         Level

        d1,d2,d3                      d1,d2,d3                     3
                                      /  \
                                     /    \
   d1,d2   d2,d3   d1,d3       d1,d2        d2,d3                   2
                                     \    /
                                      \  /
     d1      d2      d3                d2                           1
```

The set **d1,d2,d3** corresponds to $d(b,e)$ and $d(e,f)$
The subset **d1,d2** corresponds to $d(a,b)$, $d(a,f)$, $d(b,c)$, $d(b,f)$ and $d(c,f)$
The subset **d2,d3** corresponds to $d(a,e)$ and $d(c,e)$
The subset **d2** corresponds to $d(a,c)$

Clusters defined by all pairwise linkage at level ≤ 2:
a,b,c,f
a,c,e

Clusters defined by all pairwise linkage at level ≤ 3:
a,b,c,e,f

FIGURE 3.3: Top: example data set consisting of five objects, characterized by three boolean attributes. See the description in the text of how similarity is a vector of three components, associated with the variables here. Then there is a small depiction of the lattice corresponding to this data set, followed by its specification.

$d : I \times I \to 2^J$. The latter, leading to clustering based on dissimilarities, was developed initially in [101].

3.4.2   Applications of Generalized Ultrametrics

As already noted, the usual ultrametric is an ultrametric distance, that is, for a set I, $d : I \times I \longrightarrow \mathbb{R}^+$. The generalized ultrametric is $d : I \times I \longrightarrow \Gamma$, where Γ is a partially ordered set. In other words, the *generalized* ultrametric distance is a set. Some areas of application of generalized ultrametrics will now be discussed.

In the theory of reasoning, a monotonic operator is rigorous application of a succession of conditionals (sometimes called consequence relations). However, negation or multiple-valued logic (i.e. encompassing intermediate truth and falsehood) require support for non-monotonic reasoning.

Thus [96]: "Once one introduces negation ... then certain of the important operators are not monotonic (and therefore not continuous), and in consequence the Knaster–Tarski theorem [i.e. for fixed points; see [56]] is no longer applicable to them. Various ways have been proposed to overcome this problem. One such [approach is to use] syntactic conditions on programs ... Another is to consider different operators ... The third main solution is to introduce techniques from topology and analysis to augment arguments based on order ... [the latter include:] methods based on metrics ... on quasi-metrics ... and finally ... on ultrametric spaces."

The convergence to fixed points that are based on a generalized ultrametric system is precisely the study of spherically complete systems and expansive automorphisms discussed in Section 3.5.3 below. As expansive automorphisms we see here again an example of symmetry at work.

A direct application of generalized ultrametrics to data mining is the following. The potentially huge advantage of the generalized ultrametric is that it allows a hierarchy to be read directly off the $I \times J$ input data, and bypasses the $O(n^2)$ consideration of all pairwise distances in agglomerative hierarchical clustering. In [186] there is the study of application to chemoinformatics; for this application, see Section 6.11 below. Proximity and best match finding is an essential operation in this field. Typically we have 1 million chemicals upwards, characterized by an approximate 1000-valued attribute encoding.

3.5   Hierarchy in a p-Adic Number System

A dendrogram is widely used in hierarchical, agglomerative clustering, and is induced from observed data. In this chapter, one of our important goals is to show how it lays bare many diverse symmetries in the observed phenomenon represented by the data. By expressing a dendrogram in p-adic terms, we open up a wide range of possibilities for seeing symmetries and attendant invariants.

3.5.1 p-Adic Encoding of a Dendrogram

We will now introduce the one-to-one mapping of clusters (including singletons) in a dendrogram H into a set of p-adically expressed integers (*a forteriori*, rationals or $\mathbb{Q}_p$). The field of p-adic numbers is the most important example of an ultrametric space. Addition and multiplication of p-adic integers, $\mathbb{Z}_p$ (cf. the expression for a positive integer in Section 3.2.4), are well defined. Inverses exist and no zero-divisors exist.

A terminal-to-root traversal in a dendrogram or binary rooted tree is defined as follows. We use the path $x \subset q \subset q' \subset q'' \subset \cdots \subset q_{n-1}$, where x is a given object specifying a given terminal, and $q, q', q'', \ldots$ are the embedded classes along this path, specifying nodes in the dendrogram. The root node is specified by the class q_{n-1} comprising all objects. A terminal-to-root traversal is the shortest path between the given terminal node and the root node, assuming we preclude repeated traversal (backtrack) of the same path between any two nodes.

By means of terminal-to-root traversals, we define the following p-adic encoding of terminal nodes, and hence objects, in Figure 3.4:

$$
\begin{aligned}
x_1 : \quad & +1 \cdot p^1 + 1 \cdot p^2 + 1 \cdot p^5 + 1 \cdot p^7 & (3.1) \\
x_2 : \quad & -1 \cdot p^1 + 1 \cdot p^2 + 1 \cdot p^5 + 1 \cdot p^7 \\
x_3 : \quad & -1 \cdot p^2 + 1 \cdot p^5 + 1 \cdot p^7 \\
x_4 : \quad & +1 \cdot p^3 + 1 \cdot p^4 - 1 \cdot p^5 + 1 \cdot p^7 \\
x_5 : \quad & -1 \cdot p^3 + 1 \cdot p^4 - 1 \cdot p^5 + 1 \cdot p^7 \\
x_6 : \quad & -1 \cdot p^4 - 1 \cdot p^5 + 1 \cdot p^7 \\
x_7 : \quad & +1 \cdot p^6 - 1 \cdot p^7 \\
x_8 : \quad & -1 \cdot p^6 - 1 \cdot p^7
\end{aligned}
$$

If we choose $p = 2$ the resulting decimal equivalents could be the same: compare contributions based on $+1 \cdot p^1$ and $-1 \cdot p^1 + 1 \cdot p^2$. Given that the coefficients of the p^j terms $(1 \leq j \leq 7)$ are in the set $\{-1, 0, +1\}$ (implying for x_1 the additional terms $+0 \cdot p^3 + 0 \cdot p^4 + 0 \cdot p^6$), the coding based on $p = 3$ is required to avoid ambiguity among decimal equivalents.

A few general remarks on this encoding follow. For the labelled ranked binary trees that we are considering, we require the labels $+1$ and -1 for the two branches at any node. Of course we could interchange these labels, and have these $+1$ and -1 labels reversed at any node. By doing so we will have different p-adic codes for the objects, x_i.

The following properties hold: (i) the decimal codes for each x_i (lexicographically ordered) are unique for $p \geq 3$ (*unique encoding*); and (ii) the dendrogram can be uniquely reconstructed from any such set of unique codes (*reversibility*).

The p-adic encoding defined for any object set can be expressed as follows for any object x associated with a terminal node:

$$
x = \sum_{j=1}^{n-1} c_j p^j, \quad \text{where } c_j \in \{-1, 0, +1\}. \tag{3.2}
$$

In greater detail, we have

$$
x_i = \sum_{j=1}^{n-1} c_{ij} p^j, \quad \text{where } c_{ij} \in \{-1, 0, +1\}. \tag{3.3}
$$

Here j is the level or rank (root, $n-1$; terminal, 1), and i is an object index.

In our example we have used $c_j = +1$ for a left branch (in the sense of Figure 3.4), -1 for a right branch, and 0 when the node is not on the path from that particular terminal to the root.

A matrix form of this encoding is as follows, where $\{\cdot\}^t$ denotes the transpose of the vector. Let $\mathbf{x}$ be the column vector $\{x_1 \ x_2 \ \ldots \ x_n\}^t$. Let $\mathbf{p}$ be the column vector $\{p^1 \ p^2 \ \ldots \ p^{n-1}\}^t$. Define a characteristic matrix C of the branching codes, $+1$ and -1, and an absent or non-existent branching given by 0, as a set of values c_{ij} where $i \in I$, the

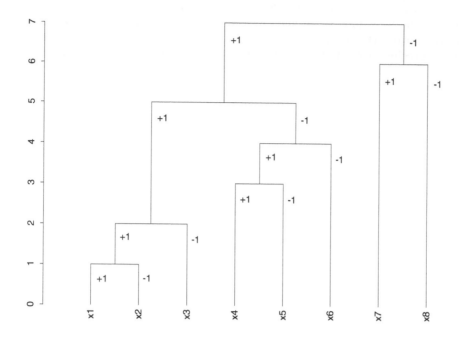

FIGURE 3.4: Labelled, ranked dendrogram on eight terminal nodes, $x_1, x_2, \ldots, x_8$. Branches are labelled $+1$ and -1. Clusters are $q_1 = \{x_1, x_2\}, q_2 = \{x_1, x_2, x_3\}, q_3 = \{x_4, x_5\}, q_4 = \{x_4, x_5, x_6\}, q_5 = \{x_1, x_2, x_3, x_4, x_5, x_6\}, q_6 = \{x_7, x_8\}, q_7 = \{x_1, x_2, \ldots, x_7, x_8\}$.

indices of the object set, and $j \in \{1, 2, \ldots, n-1\}$, the indices of the dendrogram levels or nodes ordered increasingly. For Figure 3.4 we therefore have

$$C = \{c_{ij}\} = \begin{pmatrix} 1 & 1 & 0 & 0 & 1 & 0 & 1 \\ -1 & 1 & 0 & 0 & 1 & 0 & 1 \\ 0 & -1 & 0 & 0 & 1 & 0 & 1 \\ 0 & 0 & 1 & 1 & -1 & 0 & 1 \\ 0 & 0 & -1 & 1 & -1 & 0 & 1 \\ 0 & 0 & 0 & -1 & -1 & 0 & 1 \\ 0 & 0 & 0 & 0 & 0 & 1 & -1 \\ 0 & 0 & 0 & 0 & 0 & -1 & -1 \end{pmatrix} . \tag{3.4}$$

For given level j, for all i, the absolute values $|c_{ij}|$ give the membership function either by node, j, which is therefore read off columnwise; or by object index, i, which is therefore read off rowwise.

The matrix form of the p-adic encoding used in equations (3.2) or (3.3) is

$$\mathbf{x} = C\mathbf{p}. \tag{3.5}$$

Here, $\mathbf{x}$ is the decimal encoding, C is the matrix with dendrogram branching codes (cf. the example shown in expression (3.4)), and $\mathbf{p}$ is the vector of powers of a fixed integer (usually, more restrictively, fixed prime) p.

The tree encoding exemplified in Figure 3.4, and defined with coefficients in equations

(3.2) or (3.3), (3.4) or (3.5), with labels $+1$ and -1 (as opposed to the choice of 0 and 1, which might have been our first thought), was required to fully cater for the ranked nodes (i.e. the total order, as opposed to a partial order, on the nodes).

We can consider the objects that we are dealing with to have equivalent integer values. To show that, all we must do is work out decimal equivalents of the p-adic expressions used above for $x_1, x_2, \ldots$. As noted in [90], we have equivalence between a p-adic number, a p-adic expansion and an element of $\mathbb{Z}_p$ (the p-adic integers). The coefficients used to specify a p-adic number, [90, p. 69] notes, "must be taken in a set of representatives of the class modulo p. The numbers between 0 and $p - 1$ are only the most obvious choice for these representatives. There are situations, however, where other choices are expedient."

We note that the matrix C is used in [53]. A somewhat trivial view of how "hierarchical trees can be perfectly scaled in one dimension" (the title and theme of [53]) is that p-adic numbering is feasible, and hence a one-dimensional representation of terminal nodes is easily arranged through expressing each p-adic number with a real number equivalent.

3.5.2   *p*-Adic Distance on a Dendrogram

We will now induce a metric topology on the p-adically encoded dendrogram, H. It leads to various symmetries relative to identical norms, for instance, or identical tree distances. For convenience, we will use a *similarity* which we can convert to a distance.

To find the p-adic similarity, we look for the term p^r in the p-adic codes of the two objects, where r is the lowest level such that the values of the coefficients of p^r differ.

Let us look at the set of p-adic codes for $x_1, x_2, \ldots$ above (Figure 3.4 and relations (3.2)), to give some examples of this.

For x_1 and x_2, we find the term we are looking for to be p^1, and so $r = 1$.
For x_1 and x_5, we find the term we are looking for to be p^5, and so $r = 5$.
For x_5 and x_8, we find the term we are looking for to be p^7, and so $r = 7$.

Having found the value r, the similarity is defined as p^{-r} [25, 90].

We take for a singleton object $r = 0$, and so a similarity s has the property that $s(x, y) \leq 1, x \neq y$, and $s(x, x) = 1$. This leads naturally to an associated distance $d(x, y) = 1 - s(x, y)$, which is furthermore a 1-bounded ultrametric.

An alternative way of looking at the p-adic similarity (or distance) introduced, from the p-adic expansions listed in relations (3.2), is as follows. Consider the longest common sequence of coefficients using terms of the expansion from the start of the sequence. We will ensure that the start of the sequence corresponds to the root of the tree representation. Determine the p^r term before which the value of the coefficients first differ. Then the similarity is defined as p^{-r} and distance as $1 - p^{-r}$.

This longest common prefix metric is also known as the Baire distance. In topology the Baire metric is defined on infinite strings [139]. It is more than just a distance: it is an ultrametric bounded from above by 1, and its infimum is 0, which is relevant for very long sequences, or in the limit for infinite-length sequences. The use of this Baire metric is pursued in [186] based on random projections [238], and providing computational benefits over the classical $O(n^2)$ hierarchical clustering based on all pairwise distances.

The longest common prefix metric leads directly to a *p-adic hierarchical classification* (cf. [35]). This is a special case of the "fast" hierarchical clustering discussed at the end of Section 3.4.2.

Compared to the longest common prefix metric, there are other closely related forms of metric, and simultaneously ultrametric. In [84], the metric is defined via the integer part of a real number. In [25], for integers x, y we have $d(x, y) = 2^{-\text{order}_p(x-y)}$ where p is prime, and $\text{order}_p(i)$ is the exponent (non-negative integer) of p in the prime decomposition of an integer. Furthermore, let $S(x)$ be a series: $S(x) = \sum_{i \in \mathbb{N}} a_i x^i$ ($\mathbb{N}$ are the natural numbers). The order of $S(i)$ is the rank of its first non-zero term: $\text{order}(S) = \inf\{i : i \in \mathbb{N}; a_i \neq 0\}$. (The series that is all zero is of order infinity.) Then the ultrametric similarity between series is $d(S, S') = 2^{-\text{order}(S-S')}$.

3.5.3   Scale-Related Symmetry

Scale-related symmetry is very important in practice. In this subsection we introduce an operator that provides this symmetry. We also term it a dilation operator, because of its role in the wavelet transform on trees (see [173] for discussion and examples). This operator is p-adic multiplication by $1/p$.

Consider the set of objects $\{x_i | i \in I\}$ with its p-adic coding considered above. Take $p = 2$. (Non-uniqueness of corresponding decimal codes is not of concern to us now, and taking this value for p is without any loss of generality.) Multiplication of $x_1 = +1 \cdot 2^1 + 1 \cdot 2^2 + 1 \cdot 2^5 + 1 \cdot 2^7$ by $1/p = 1/2$ gives: $+1 \cdot 2^1 + 1 \cdot 2^4 + 1 \cdot 2^6$. Each level has decreased by 1, and the lowest level has been lost. Subject to the lowest level of the tree being lost, the form of the tree remains the same. By carrying out the multiplication-by-$1/p$ operation on all objects, it is seen that the effect is to rise in the hierarchy by one level.

Let us call product with $1/p$ the operator A. The effect of losing the bottom level of the dendrogram means that either (i) each cluster (possibly singleton) remains the same, or (ii) two clusters are merged. Therefore the application of A to all q implies a subset relationship between the set of clusters $\{q\}$ and the result of applying A, $\{Aq\}$.

Repeated application of the operator A gives $Aq, A^2q, A^3q, \ldots$. Starting with any singleton, $i \in I$, this gives a path from the terminal to the root node in the tree. Each such path ends with the null element, which we define to be the p-adic encoding corresponding to the root node of the tree. Therefore the intersection of the paths equals the null element.

Benedetto and Benedetto [21, 22] discuss A as an expansive automorphism of I, that is, form-preserving and locally expansive. Some implications [21] of the expansive automorphism follow. For any q, let us take $q, Aq, A^2q, \ldots$ as a sequence of open subgroups of I, with $q \subset Aq \subset A^2q \subset \ldots$, and $I = \bigcup\{q, Aq, A^2q, \ldots\}$. This is termed an inductive sequence of I, and I itself is the inductive limit [210, p. 131].

Each path defined by application of the expansive automorphism defines a spherically complete system [212, 84, 237], which is a formalization of well-defined subset embeddedness. Such a methodological framework finds application in multi-valued and non-monotonic reasoning, as noted in Section 3.4.2.

3.6   Tree Symmetries through the Wreath Product Group

In this section the wreath product group, used up to now in the literature as a framework for tree structuring of image or other signal data, is here used on a two-way tree or dendrogram data structure. An example of wreath product invariance is provided by the wavelet transform of such a tree.

3.6.1   Wreath Product Group for Hierarchical Clustering

A dendrogram like that shown in Figure 3.4 is invariant as a representation or structuring of a data set relative to rotation (alternatively here, permutation) of left and right child nodes. These rotation (or permutation) symmetries are defined by the wreath product group (see

[79, 80, 77] for an introduction and applications in signal and image processing), and can be used with any m-ary tree, although we will treat the binary or two-way case here.

For the group actions, with respect to which we will seek invariance, we consider independent cyclic shifts of the subnodes of a given node (hence, at each level). Equivalently, these actions are adjacency-preserving permutations of subnodes of a given node (i.e. for given q, with $q = q' \cup q''$, the permutations of $\{q', q''\}$). We have therefore cyclic group actions at each node, where the cyclic group is of order 2.

The symmetries of H are given by structured permutations of the terminals. The terminals will be denoted here by Term H. The full group of symmetries is summarized by the following generative algorithm:

1. For level $l = n - 1$ down to 1 do:

2. Selected node, $\nu \longleftarrow$ node at level l.

3. And permute subnodes of ν.

Subnode ν is the root of subtree H_ν. We denote H_{n-1} simply by H. For a subnode ν' undergoing a relocation action in step 3, the internal structure of subtree $H_{\nu'}$ is not altered.

The algorithm described defines the automorphism group which is a wreath product of the symmetric group. Denote the permutation at level ν by P_ν. Then the automorphism group is given by

$$G = P_{n-1} \text{ wr } P_{n-2} \text{ wr } \ldots \text{ wr } P_2 \text{ wr } P_1,$$

where wr denotes the wreath product.

3.6.2 Wreath Product Invariance

Call Term H_ν the terminals that descend from the node at level ν. So these are the terminals of the subtree H_ν with its root node at level ν. We can alternatively call Term H_ν the cluster associated with level ν.

We will now look at shift invariance under the group action. This amounts to the requirement for a constant function defined on Term H_ν for all ν. A convenient way to do this is to define such a function on the set Term H_ν via the root node alone, ν. By definition then we have a constant function on the set Term H_ν.

Let us call V_ν a space of functions that are constant on Term H_ν. That is to say, the functions are constant in clusters that are defined by the subset of n objects. Possibilities for V_ν that were considered in [173] are:

(i) basis vector with |Term H_{n-1}| components, with 0 values except for value 1 for component i;

(ii) set (of cardinality $n = |$Term $H_{n-1}|$) of m-dimensional observation vectors.

Consider the resolution scheme arising from moving from {Term $H_{\nu'}$, Term $H_{\nu''}$} to Term H_ν. From the hierarchical clustering point of view it is clear what this represents: simply, an agglomeration of two clusters called Term $H_{\nu'}$ and Term $H_{\nu''}$, replacing them with a new cluster, Term H_ν.

Let the spaces of functions that are constant on subsets corresponding to the two cluster agglomerands be denoted $V_{\nu'}$ and $V_{\nu''}$. These two clusters are disjoint initially, which motivates us taking the two spaces as a couple: $(V_{\nu'}, V_{\nu''})$.

	Sepal.L	Sepal.W	Petal.L	Petal.W
1	5.1	3.5	1.4	0.2
2	4.9	3.0	1.4	0.2
3	4.7	3.2	1.3	0.2
4	4.6	3.1	1.5	0.2
5	5.0	3.6	1.4	0.2
6	5.4	3.9	1.7	0.4
7	4.6	3.4	1.4	0.3
8	5.0	3.4	1.5	0.2

TABLE 3.3: The first eight observations of Fisher's iris data. L and W refer to length and width.

3.6.3 Wreath Product Invariance: Haar Wavelet Transform of Dendrogram

Let us exemplify a case that satisfies all that has been defined in the context of the wreath product invariance that we are targeting. It is the algorithm discussed in detail in [173]. Take the constant function from $V_{\nu'}$ to be $f_{\nu'}$. Take the constant function from $V_{\nu''}$ to be $f_{\nu''}$. Then define the constant function, the *scaling function*, in V_ν to be $(f_{\nu'} + f_{\nu''})/2$. Next define the zero-mean function, $(w_{\nu'} + w_{\nu''})/2 = 0$, the *wavelet function*, as follows:

$$w_{\nu'} = (f_{\nu'} + f_{\nu''})/2 - f_{\nu'}$$

in the support interval of $V_{\nu'}$ (i.e. Term $H_{\nu'}$), and

$$w_{\nu''} = (f_{\nu'} + f_{\nu''})/2 - f_{\nu''}$$

in the support interval of $V_{\nu''}$ (i.e. Term $H_{\nu''}$). Since $w_{\nu'} = -w_{\nu''}$ we have the zero mean requirement.

We now illustrate the Haar wavelet transform of a dendrogram with a case study.

The discrete wavelet transform is a decomposition of data into spatial and frequency components. In terms of a dendrogram these components are with respect to, respectively, within and between clusters of successive partitions. We show how this works taking the data of Table 3.3.

The hierarchy built on the eight observations of Table 3.3 is shown in Figure 3.5. Here we label irises 1 to 8 as, respectively, $x_1, x_3, x_4, x_6, x_8, x_2, x_5, x_7$.

Something more is shown in Figure 3.5, namely the detail signals (denoted $\pm d$) and overall smooth (denoted s), which are determined in carrying out the wavelet transform, the so-called forward transform.

The inverse transform is then determined from Figure 3.5 in the following way. Consider the observation vector x_2. This vector is reconstructed exactly by reading the tree from the root: $s_7 + d_7 = x_2$. Similarly, a path from root to terminal is used to reconstruct any other observation. If x_2 is a vector of dimensionality m, then so also are s_7 and d_7, as well as all other detail signals.

This procedure is the same as the Haar wavelet transform, only applied to the dendrogram and using the input data. This wavelet transform for the data in Table 3.3, based on the "key" or intermediary hierarchy of Figure 3.5, is shown in Table 3.4.

Wavelet regression entails setting small and hence unimportant detail coefficients to 0 before applying the inverse wavelet transform. More discussion can be found in [173].

Early work on p-adic and ultrametric wavelets can be found in Kozyrev [124, 125]. Recent applications of wavelets to general graphs are in [193, 103].

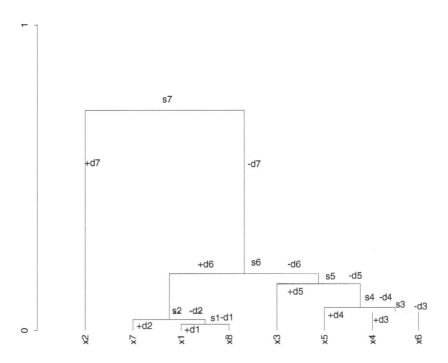

FIGURE 3.5: Dendrogram on eight terminal nodes constructed from the first eight values of Fisher's iris data. (Median agglomerative method used in this case.) Detail or wavelet coefficients are denoted by d, and data smooths are denoted by s. The observation vectors are denoted by x and are associated with the terminal nodes. Each *signal smooth*, s, is a vector. The (positive or negative) *detail signals*, d, are also vectors. All these vectors are of the same dimensionality.

	s_7	d_7	d_6	d_5	d_4	d_3	d_2	d_1
Sepal.L	5.146875	0.253125	0.13125	0.1375	−0.025	0.05	−0.025	0.05
Sepal.W	3.603125	0.296875	0.16875	−0.1375	0.125	0.05	−0.075	−0.05
Petal.L	1.562500	0.137500	0.02500	0.0000	0.000	−0.10	0.050	0.00
Petal.W	0.306250	0.093750	−0.01250	−0.0250	0.050	0.00	0.000	0.00

TABLE 3.4: The hierarchical Haar wavelet transform resulting from use of the first eight observations of Fisher's iris data shown in Table 3.3. Wavelet coefficient levels are denoted $d_1, \ldots, d_7$, and the continuum or smooth component is denoted s_7.

3.7   Tree and Data Stream Symmetries from Permutation Groups

In this section we show how data streams, firstly, and hierarchies, secondly, can be represented as permutations. There are restrictions on permitted permutations. Furthermore, sets of data streams, or of trees, when expressed as permutations constitute particular permutation groups.

3.7.1   Permutation Representation of a Data Stream

In symbolic dynamics, we seek to extract symmetries in the data based on topology alone, before considering metric properties. For example, instead of listing a sequence of iterates, $\{x_i\}$, we may symbolically encode the sequence in terms of up and down, or north, south, east and west, moves. This provides a sequence of symbols, and their patterns in a phase space, where the interest of the data analyst lies in a partition of the phase space. Patterns or templates are sought in this topology. Sequence analysis is tantamount to a sort of topological time series analysis.

Thus, in symbolic dynamics, the data values in a stream or sequence are replaced by symbols to facilitate pattern-finding, in the first instance, through topology of the symbol sequence. This can be very helpful for analysis of a range of dynamical systems, including chaotic, stochastic, and deterministic-regular time series. Through measure-theoretic or Kolmogorov–Sinai entropy of the dynamical system, it can be shown that the maximum entropy conditional on past values is consistent with the requirement that the symbol sequence retains as much of the original data information as possible. Alternative approaches to quantifying complexity of the data, expressing the dynamical system, are through Lyapunov exponents and fractal dimensions, and there are close relationships between all of these approaches [128].

From the viewpoint of practical and real-world data analysis, however, many problems and open issues remain. Firstly, noise in the data stream means that reproducibility of results can break down [15]. Secondly, the symbol sequence, and derived partitions that are the basis for the study of the symbolic dynamic topology, are not easy to determine. Hence Bandt and Pompe [15] enunciate a pragmatic principle whereby the symbol sequence should come as naturally as possible from the data, with as little as possible by way of further model assumptions. Their approach is to define the symbol sequence through (i) comparison of neighbouring data values, and (ii) up–down or down–up movements in the data stream. Taking into account all up–down and down–up movements in a signal allows a permutation representation.

Examples of such symbol sequences from Bandt and Pompe [15] follow. They consider the data stream $(x_1, x_2, \ldots, x_7) = (4, 7, 9, 10, 6, 11, 3)$. Take the order as 3, that is, consider the up–down and down–up properties of successive triplets: $(4, 7, 9) \longrightarrow 012; (7, 9, 10) \longrightarrow 012; (9, 10, 6) \longrightarrow 201; (6, 11, 3) \longrightarrow 201; (10, 6, 11) \longrightarrow 102$. (In the last, for instance, we have $x_{t+1} < x_t < x_{t+2}$, yielding the symbolic sequence 102. In the second last, there is the 3rd, 1st, 2nd in value, hence 201.) In addition to the order, here 3, we may also consider the delay, here 1. In general, for delay τ, the neighbourhood consists of data values indexed by $t, t - \tau, t - 2\tau, t - 3\tau, \ldots, t - d\tau$, where d is the order. Thus, in the example used here, we have the symbolic representation from the succession of triplets in the given data stream, 012 012 201 102 201. The symbol sequence (or "itinerary") defines a partition – a separation of phase space into disjoint regions (here, with three equivalence classes, 012, 201, and 102), which facilitates finding an "organizing template" or set of topological relationships [242]. The problem is described in [109] as one of studying the qualitative behaviour of the dynamical system, through use of a "very coarse-grained" description, which divides the state space (or phase space) into a small number of regions, and codes each by a different symbol.

Different encodings are feasible and Keller and Sinn [112, 111] use the following. Again consider the data stream $(x_1, x_2, \ldots, x_7) = (4, 7, 9, 10, 6, 11, 3)$. Given a delay $\tau = 1$, we can represent the above by $(x_{6\tau}, x_{5\tau}, x_{4\tau}, x_{3\tau}, x_{2\tau}, x_\tau, x_0)$. Now look at rank order and note that $x_\tau > x_{3\tau} > x_{4\tau} > x_{5\tau} > x_{2\tau} > x_{6\tau} > x_0$. We read off the final permutation representation as (1345260). There are many ways of defining such a permutation, none of them best, asKeller and Sinn [112] acknowledge. We see too that our m-valued input stream is a point in $\mathbb{R}^m$, and our output is a permutation $\pi \in S_m$, that is, a member of the permutation group.

Keller and Sinn [112] explore invariance properties of the permutations expressing the ordinal, symbolic coding. Resolution scale is introduced through the delay, τ. (An alternative approach to incorporating resolution scale is used in [52], where consecutive, sliding-window based, binned or averaged versions of the time series are used. This is not entirely satisfactory: it is not robust and is very dependent on data properties such as dynamic range.) Application is to EEG (univariate) signals (with some discussion of magnetic resonance imaging data) [110]. Statistical properties of the ordinal transformed data are studied in [16], in particular through the S_3 symmetry group. We have noted the symbolic dynamics motivation for this work; in [14] and other work, motivation is provided in terms of rank-order time series analysis, in turn motivated by the need for robustness in time series data analysis.

3.7.2 Permutation Representation of a Hierarchy

There is an isomorphism between the class of hierarchic structures, termed unlabelled, ranked, binary, rooted trees, and the class of permutations used in symbolic dynamics. Each non-terminal node in the tree shown in Figure 3.6 has two child nodes. This is a dendrogram, representing a set of $n - 1$ agglomerations based on n initial data vectors.

Figure 3.6 shows a hierarchical clustering. Figure 3.7 shows a unique representation of the tree, termed a dendrogram, subject only to terminals being permutable in position relative to the first non-terminal cluster node.

A *packed representation* [217] or permutation representation of a dendrogram is derived as follows. Put a lower-ranked subtree always to the left, and read off the oriented binary tree on non-terminal nodes. Then for any terminal node indexed by i, with the exception of the rightmost which will always be n, define $p(i)$ as the rank at which the terminal node is first united with some terminal node to its right.

For the dendrogram shown in Figure 3.9 (or Figure 3.7 or 3.8), the packed representation is (13625748). This is also an inorder traversal of the oriented binary tree. The packed representation is a uniquely defined permutation of $1 \ldots n$.

Dendrograms (on n terminals) of the sort shown in Figures 3.6–3.9 are labelled (see terminal node labels, "existence", "object", etc.) and ranked (ranks indicated in Figure 3.8). Consider when tree structure alone is of interest and we ignore the labels. Such dendrograms, called unlabelled, ranked (NL-R) in [165], are particularly interesting. They are isomorphic to either down–up permutations or up–down permutations (both on $n - 1$ elements). For the combinatorial properties of these permutations, and NL-R dendrograms, see the combinatorial sequence encyclopedia entry, A000111, related to the sequences termed André numbers and Euler numbers, at [220].

We see therefore how we are dealing with the group of up–down or down–up permutations.

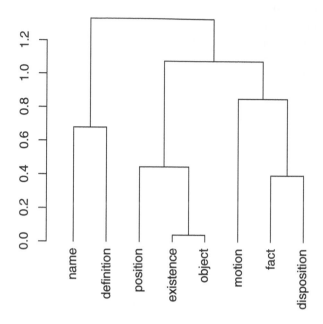

FIGURE 3.6: Hierarchical clustering of eight terms. The data on which this was based are frequencies of occurrence of the eight nouns in 24 successive, non-overlapping segments of Aristotle's *Categories*.

3.8 Remarkable Symmetries in Very High-Dimensional Spaces

In the work of [206, 207] it was shown how as ambient dimensionality increased distances became more and more ultrametric. That is to say, a hierarchical embedding becomes more and more immediate and direct as dimensionality increases. A better way of quantifying this phenomenon was developed in [169]. What this means is that there is inherent hierarchical structure in high-dimensional data spaces.

It was shown experimentally in [206, 207, 169] how points in high-dimensional spaces become increasingly equidistant with increase in dimensionality. Both [92] and [65] study Gaussian clouds in very high dimensions. The latter finds that "not only are the points [of a Gaussian cloud in very high-dimensional space] on the convex hull, but all reasonable-sized subsets span faces of the convex hull. This is wildly different than the behavior that would be expected by traditional low-dimensional thinking."

That very simple structures come about in very high dimensions is not as trivial as it might appear at first sight. Firstly, even very simple structures (hence with many symmetries) can be used to support fast and perhaps even constant-time worst-case proximity search [169]. Secondly, as shown in the machine learning framework by [92], there are important implications ensuing from the simple high-dimensional structures. Thirdly, [175] shows that very high-dimensional clustered data contain symmetries that in fact can be exploited

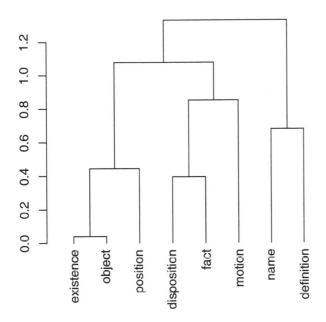

FIGURE 3.7: Dendrogram on eight terms, isomorphic to Figure 3.6, but now with successively *later* agglomerations always represented by *right* child nodes. Apart from the labels of the initial pairwise agglomerations, this is otherwise a unique representation of the dendrogram (hence, "existence" and "object" can be interchanged; so can "disposition" and "fact"; and finally, "name" and "disposition"). In the discussion we refer to this representation, with later agglomerations always parked to the right, as our canonical representation of the dendrogram.

to "read off" the clusters in a computationally efficient way. Fourthly, following [58], what we might want to look for in contexts of considerable symmetry are the "impurities" or small irregularities that detract from the overall dominant picture.

3.9   Short Commentary on This Chapter

"My thesis has been that one path to the construction of a nontrivial theory of complex systems is by way of a theory of hierarchy" [219, p. 216]. Or again: "Human thinking (as well as many other information processes) is fundamentally a hierarchical process. ... In our information modeling the main distinguishing feature of p-adic numbers is the treelike hierarchical structure. ... [the work] is devoted to classical and quantum models of flows of hierarchically ordered information" [115, pp. xiii, xv].

We have noted symmetry in many guises in the representations used, in the transformations applied, and in the transformed outputs. These symmetries are non-trivial too, in

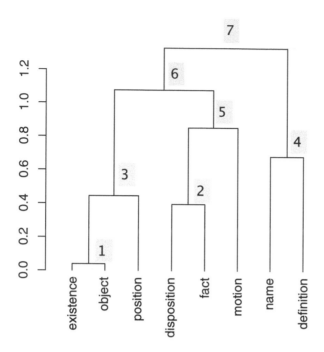

FIGURE 3.8: Dendrogram on eight terms, as Figure 3.7, with non-terminal nodes numbered in sequence. These will form the nodes of the oriented binary tree. We may consider one further node for completeness, 8 or ∞, located at an arbitrary location in the upper right.

a way that would not be the case were we simply to look at classes of a partition and claim that cluster members were mutually similar in some way. We have seen how the p-adic or ultrametric framework provides significant focus and commonality of viewpoint.

In seeking (in a general way) and in determining (in a focused way) structure and regularity in data, we see that, in line with the insights and achievements of Klein, Weyl and Wigner, in data mining and data analysis we seek and determine symmetries in the data that express observed and measured reality. A very fundamental principle in much of statistics, signal processing and data analysis is that of sparsity, but, as shown in [17], by "codifying the inter-dependency structure" in the data new perspectives are opened up above and beyond sparsity.

Major applications of what has been covered in this chapter will be central to the content of Chapters 8 and 9.

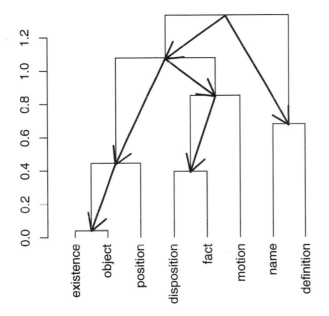

FIGURE 3.9: Oriented binary tree superimposed on the dendrogram. The node at the arbitrary upper right location is not shown. The oriented binary tree defines an inorder or depth-first tree traversal.

4

Geometry and Topology of Data Analysis: in p-Adic Terms

The data analysis framework, that is, the geometry and topology of the data that we are both investigating and making use of, will be considered in this chapter in p-adic number system terms, rather than in real number terms. By notational convention, we use p for a prime number, and m for some non-prime integer. That prime number systems are all-pervasive for us is a reasonable view, given the binary number system, $p = 2$, that computers use, and when we recall that Nikolay Brusentsov's development of computers in the 1950s and 1960s used a ternary number system, $p = 3$. While the decimal number system, $m = 10$, is so dominant, we still continue to use other number systems such as the sexagesimal, $m = 60$, for time.

Here we take further our mathematical underpinnings, in particular in relation to ultrametric (i.e. hierarchical) topology. In mathematics a completely alternative approach relative to the real numbers is provided by p-adic number theory. We provide a broad and general overview of this. We discuss some future perspectives for this area in cosmology, in genomics, and for complex systems generally.

Section 4.1 describes, with examples, what p-adic numbers are all about. They are an alternative to the real numbers. By extension, the real numbers are related to the more general context of complex numbers. The reals are central to vector spaces that are specified in the field of the reals. There is a limitation there, however, in regard to measurements and quantification of real phenomena. In this chapter we will recount how one such limitation is discreteness at very small physical scales. Another limitation is a major motivation for us. This limitation is that real numbers are impractical, and of questionable usefulness, on physical scales that may become unboundedly big. The alternative p-adic number representation handles spatial scales well that are enormously large and very small.

Benzécri [25, p. 140] gives a short discussion of p-adic numbers and p-adic, or ultrametric, distance.

4.1 Numbers and Their Representations

First, some scene-setting. The most fundamental numbers are the natural numbers, $\mathbb{N} = 0, 1, 2, 3, \ldots$, and the rational numbers, $\mathbb{Q} = a/b$, where $a, b \in \mathbb{N}$ and where we avoid trouble by requiring that $b \neq 0$. The integers, $\mathbb{Z}$, are $\ldots, -3, -2, -1, 0, 1, 2, 3, \ldots$. With integers, a rational number is x/y, where $x, y \in \mathbb{Z}$.

4.1.1 Series Representations of Numbers

A positive real number, a, can be written as an infinite decimal fraction, $a = \sum_{k=m}^{\infty} a_k 10^{-k}$, where m is an integer (positive or negative or 0), and the coefficients are integers such that:

$0 \leq a_k \leq 9$ (that is, $a \in \{0, 1, 2, 3, 4, 5, 6, 7, 8, 9\}$). For example, $a = 25.333333\ldots$ means the following.

For $m = -1$, $a_{-1} = 2$, i.e. the coefficient of $10^{-(-1)} = 10$.

For $m = 0$, $a_0 = 5$, i.e. the coefficient of $10^0 = 1$.

For $m = 1$, $a_1 = 3$, i.e. the coefficient of $10^{-1} = 0.1$.

For $m = 2$, $a_2 = 3$, i.e. the coefficient of $10^{-2} = 0.01$.

And so on.

This representation is unique unless $a_k = 0$ for all $k > n$ and $a_n \neq 0$. An example of when this is 4.7804, where we have $a_k = 0$ for all $k > 5$. We see that $n = 4$ and $a_4 \neq 0$. We have $a_4 = 4$, the fourth decimal digit.

When the representation is not unique, there is a second representation for a. This is when $a'_k = a_k$ for $k < n$, $a'_n = a_n - 1$, and $a'_k = 9$ for all $k > n$. In the example of 4.7804, what we have as the alternative representation is $4.78039999\ldots$.

A nice point about p-adic number representation that will be explored below is that there are no such non-unique cases. In terms of series expansions, if two p-adic expansions converge to the same p-adic number, then all their digits (or coefficients in the terms of the series) are the same.

In these two representations, either precisely, or through such an infinite decimal fraction, we have a point on the "real number line" or number axis. This is the perspective used for what is termed the completion of $\mathbb{Q}$, the rationals, yielding $\mathbb{R}$, the reals.

If represented as infinite decimal fractions, rational numbers are characterized by the property that they are eventually periodic. We can construct infinite number sequences that are not periodic. Therefore, that shows that there are real numbers that are not rational. (See Katok [107, p. 5] for more discussion on periodic representations.)

Here is how completion is formally treated. A point on the real number line is a (converging) Cauchy sequence. For convergence, there is a need to have a metric defined. A metric is this mapping, defining a distance, $d : X \times X \to \mathbb{R}^+$. We have elements of a set, $x, y \in X$. We can write: $d(x, y) \in \mathbb{R}^+$, where $\mathbb{R}^+$ are the non-negative reals, and $x, y \in X$. Such a distance function has the properties of symmetry, positive definiteness, and satisfies the triangle inequality. These are respectively the following: for all $x, y, z \in X$, $d(x, y) = d(y, x)$; $d(x, y) > 0$ for $x \neq y$, and $d(x, y) = 0$ for $x = y$; and $d(x, z) \leq d(x, y) + d(y, z)$.

The representation just described for decimal numbers is easily generalized to any base, g. We have $a = \sum_{k=m}^{\infty} a_k g^{-k}$. We require that g is an integer, $g \geq 2$. The coefficients are $a_k \in \{0, 1, 2, \ldots, g - 1\}$. Note how this expansion for a has smaller and smaller terms as k increases.

4.1.2 Field

A field is a set with two binary operations, usually called addition and multiplication. For each operation, there are the properties of commutativity, associativity, existence of zero, and existence of an inverse. There is also distributivity between the two operations.

Katok [107, p. 6] also mentions other algebraic structures: abelian (or commutative) group, additive group of the field, multiplicative group of the field, commutative ring.

An important property of a field is that it does not contain zero divisors, that is, $a, b \in \mathbb{F}^\times$ such that $a \cdot b = 0$, where $\mathbb{F}^\times = \mathbb{F} \setminus 0$. See (The proof, Katok [107, Exercise 9] for further discussion.

A norm is a map from a field, $\mathbb{F}$, to the non-negative reals, satisfying positive definiteness, symmetry and the triangle inequality. A trivial norm is when $\|0\| = 0$, $\|x\| = 1$ for all $x \neq 0$. That is the case for what are termed Archimedean fields, including the real numbers. If non-Archimedean fields are at issue, then the so-called strong triangle inequality needs to hold.

If p is prime, then integers modulo p form a finite field, denoted $\mathbb{Z}/p\mathbb{Z}$ or $\mathbb{F}_p$.

4.2 p-Adic Valuation, p-Adic Absolute Value, p-Adic Norm

Prime numbers, denoted p, have particular properties. Very general motivation is as follows. They lay the basis for an inherently hierarchical expression of numerical processing. This is done through, firstly, the size or measure of number that is given by a valuation measure and then the absolute value. Secondly, following directly from that, there are the concepts of distance and of norm, the latter being distance relative to 0. In the description given below, the p-adic absolute value and the p-adic norm are defined identically. Absolute values are defined on numbers, and, here with essentially an identical definition, norms relate to fields. For reals, the absolute value is defined for the real space (or number line), $\mathbb{R}$, and a norm, for example the Euclidean norm, is defined for the n-dimensional real space, $\mathbb{R}^n$.

A p-adic valuation is what the following is termed: $\mathrm{ord}_p(x)$. For $x \in \mathbb{Z}$, it is the highest power of p dividing x. Or if $x = \frac{a}{b} \in \mathbb{Q}$, with $a, b \in \mathbb{Z}$, $b \neq 0$, the p-adic valuation of x is $\mathrm{ord}_p(a) - \mathrm{ord}_p(b)$.

From this valuation, the norm is defined. The p-adic norm of $x \in \mathbb{Z}$ or $x \in \mathbb{Q}$ is $p^{-\mathrm{ord}_p(x)}$ if $x \neq 0$; and the p-adic norm of x is 0 if $x = 0$. The p-adic norm is also referred to (with this definition) as the p-adic absolute value.

Other authors use different notation (e.g. in [107, 149]), so will henceforth switch from $\mathrm{ord}_p(x)$ to $v_p(x)$.

The p-adic absolute value of a rational number r is defined to be $|r|_p = p^{-v_p(r)}$. For example:

- $|p|_p = \frac{1}{p}$ (because our given number, $p = p^1$, so we have $v_p(r) = 1$ in this case, leading to $p^{-v_p(r)} = p^{-1}$).

- $|1|_p = 1$ (because our given number, $1 = p^0$, so we have $v_p(r) = 0$ in this case. $p^{-0} = p^0 = 1$, so $p^{-v_p(r)} = 1$).

- $|2p|_p = \frac{1}{p}$ if p is odd, and $|\frac{1}{p^2}| = p^2$. For the former, p being odd means that p can only divide once into $2p$ and we cannot have a higher-order power. See how we would have a higher-order power if $p = 2$. That is the only possible, even p-adic number. So $v_p(r) = 1$, and therefore we have the stated outcome. For the latter, we are looking for the p-adic absolute value of the number p^{-2}. So, $v_p(r) = -2$, and therefore we get the stated outcome.

Summarizing the foregoing description, we have the following. If n is an integer, its p-adic valuation is the exponent of the greatest power of p that divides n. It is written $v_p(n)$. By convention, $v_p(0) = \infty$. If $r = a/b$ is a rational, its p-adic valuation is defined as $v_p(r) = v_p(a) - v_p(b)$.

Here are a few basic examples (from Madore [149]).

- The 7-adic valuation of 7 is 1. That of 14 is also 1, as are those of 21, 28, 35, 42 or 56. (Explanation: 7^1 is the highest power of 7 that divides these numbers.)

- The 7-adic valuation of 49, on the other hand, is 2, as is that of 98. (Explanation: 7^2 is the highest power dividing these.)

- The 7-adic valuation of 343 is 3. (Explanation: 7^3 divides this number.)

- The 2-adic valuation of an integer is 0 if it is odd, it is at least 1 if it is even, at least 2 if the integer is a multiple of 4, and so on.

- The 7-adic valuation of $1/7$ is -1, and so are those of $3/7$, $1/14$, $5/56$. (Explanation: taking $5/56$, we have that $7^0 = 1$ is the highest power of 7 dividing 5; 7^1 is the highest power of 7 dividing 56; so the valuation is $0 - 1$.)

- The 7-adic valuation of $1/2$ or $8/3$ is 0. (Explanation for $8/3$: 7^0 is the highest power of 7 dividing the numerator, and it is the highest power of 7 dividing the denominator; so the valuation is $0 - 0$.)

- The 7-adic valuation of $7/3$ or $14/5$ is 1. (Explanation: both numerators are divisible by 7^1, and both denominators are divisible by 7^0. So the valuation is $1 - 0$.)

- The 7-adic valuation of $48/49$ is -2. (Explanation: the highest power of 7 dividing the numerator is 0; the highest power of 7 dividing the denominator; so the valuation is $0 - 2$.)

  Here are a few basic examples from Silverman [218].

- $v_3(18) = 2$. Explanation: $3^2 = 9$ is a divisor of 18.

- $v_2(1728) = 6$. Explanation: 6 is the highest power of 2 dividing into 1728. $2^6 = 64$ is this divisor.

- $v_5\left(\frac{49}{50}\right) = -2$. Explanation: for the numerator, the divisor with the highest power of 5 is 5^0. For the denominator, the divisor with the highest power of 5 is $5^2 = 25$. So the valuation is $0 - 2$.

- $|9|_3 = \frac{1}{9}$. Explanation: first the valuation is the highest power of 3 that is a divisor of 9. This divisor is 3^2. The p-adic absolute value is $p^{-v_p(x)} = 3^{-2}$.

- $|24|_2 = \frac{1}{8}$. Explanation: 2^3 is a divisor of 24, and it is the highest power of 2 allowing this. So the p-adic absolute value is 2^{-3}.

- $|\frac{15}{28}|_7 = 7$. Explanation: for the numerator, the valuation is 0 because 7^0 is the divisor. For the denominator, the valuation is 1 because 7^1 is a divisor. So the p-adic absolute value is $7^{-(0-1)} = 7$.

  Note that in regard to the p-adic absolute value, the higher the power of p as a divisor, the larger the valuation, and then (with a negative power in the exponent) the smaller the absolute value. We can write this as $|p^n|_p \to 0$ when $n \to \infty$. This is easily seen since $|p^n|_p = p^{-n}$.

4.3 p-Adic Numbers as Series Expansions

We only note the following in order to set the scene in this section. Then we move quickly to p-adic integers, and p-adic representations of rationals. A decimal number, say 4.57341, is a series that can be written as $x = \sum_{i=-m}^{\infty} a_i 10^{-i}$. Therefore, $m = 0, a_0 = 4, a_1 = 5, a_2 = 7, a_3 = 3, a_4 = 4, a_5 = 1, a_6 = 0$, and the zero terms continue.

To start with, we take the p-adic positive integers. For every $x \in \mathbb{Z}_p$ we can write

$$x = a_0 + a_1 p + a_2 p^2 + \ldots + a_n p^n + \ldots$$
$$= \sum_{i=0}^{\infty} a_i p^i, \quad \text{with } 0 \leq a_i \leq p - 1.$$

See Gouvêa [90] for expressions that define the coefficients in this series representation, i.e. a_i, in terms of modular definitions (using value modulo p^i). The valuation, $v_p(x)$ is the index i of the first term in the sequence which has value greater than zero: $v_p(x) = \min\{i | a_i > 0\}$.

Now we turn to the field of p-adic rational numbers. For every $x \in \mathbb{Q}_p$ we can write

$$x = a_{-m} p^{-m} + a_{-m+1} p^{-m+1} + \ldots + a_0 + a_1 p + a_2 p^2 + \ldots + a_n p^n + \ldots$$
$$= \sum_{i=-m}^{\infty} a_i p^i, \quad \text{with } 0 \leq a_i \leq p - 1.$$

Assuming from our specification here that $a_{-m} > 0$, the valuation is then $v_p(x) = -m$. Recall that it is the first term in the sequence which has value greater than zero.

Let us return to some of the examples used in Section 4.2. To start, we look at series expansions and, for the valuation, find the first term that does not vanish. For $p - 7$, we are using $a = a_0 \cdot 7^0 + a_1 \cdot 7^1 + a_2 \cdot 7^2 + \ldots$. We have

$$7 = 0 \cdot 7^0 + 1 \cdot 7^1 + 0 \cdot 7^2 + \ldots \text{ (so the valuation is 1)};$$
$$14 = 0 \cdot 7^0 + 2 \cdot 7^1 + 0 \cdot 7^2 + \ldots \text{ (so the valuation is 1)};$$
$$42 = 0 \cdot 7^0 + 6 \cdot 7^1 + 0 \cdot 7^2 + \ldots \text{ (so the valuation is 1)};$$
$$56 = 0 \cdot 7^0 + 1 \cdot 7^1 + 1 \cdot 7^2 + \ldots \text{ (so the valuation is 1)}.$$

Just to explore a little further, the next two cases are in addition to the examples in the previous section:

$$57 = 1 \cdot 7^0 + 1 \cdot 7^1 + 1 \cdot 7^2 + \ldots \text{ (so the valuation is 0)};$$
$$63 = 0 \cdot 7^0 + 2 \cdot 7^1 + 2 \cdot 7^2 + \ldots \text{ (so the valuation is 1)};$$
$$49 = 0 \cdot 7^0 + 0 \cdot 7^1 + 1 \cdot 7^2 + \ldots \text{ (so the valuation is 2)};$$
$$98 = 0 \cdot 7^0 + 0 \cdot 7^1 + 2 \cdot 7^2 + \ldots \text{ (so the valuation is 2)}.$$

7-adic expansion of $1/7$: this is the 7-adic expansion of 1 minus the 7-adic expansion of 7. So we have $0 - 1$.
7-adic expansion of $5/56$: this is the 7-adic expansion of 5 minus the 7-adic expansion of 56. So we have $0 - 1$.
7-adic expansion of $7/3$: this is the 7-adic expansion of 7 minus the 7-adic expansion of 3. So we have $1 - 0$.

4.4 Canonical p-Adic Expansion; p-Adic Integer or Unit Ball

The canonical p-adic expansion of $a = \sum_{n=-m}^{\infty} d_n p^n$ is

$$a = \ldots d_n \ldots d_2 d_1 d_0 d_{-1} \ldots d_{-m}.$$

As already described in previous sections, the norm of a p-adic number is determined from the index of the first non-zero coefficient in its canonical expansion.

Relation 1. For p-adic integers, if $a = \sum_{n=0}^{\infty} d_n p^n$ with $d_n = 0$ for $0 \leq n \leq k$, $d_k \neq 0$, then $|a|_p = p^{-k}$.

Relation 2. For p-adic rationals, if $a = \sum_{n=-m}^{\infty} d_n p^n$ where $d_{-m} \neq 0$, then $|a|_p = p^m$.

The p-adic absolute value results from the valuation, $v_p(a)$, being the index of the first non-zero coefficient in the canonical expansion of a.

A p-adic integer is defined as $\mathbb{Z}_p = \{\sum_{i=0}^{\infty} a_i p^i\}$. From Relation 1, $\mathbb{Z}_p = \{a \in \mathbb{Q}_p \mid |a|_p \leq 1\}$. $\mathbb{Z}_p$ is the unit ball with centre 0, in $\mathbb{Q}_p$ (Madore [149]). We can also note that a p-adic integer is a p-adic number with non-negative valuation [149].

Consider Relation 1 with $a_0 \neq 0$. Then $k = 0$ is the first non-zero coefficient in the canonical expansion of a. So, call this a p-adic unit. $\mathbb{Z}^+ = \{x \in \mathbb{Z}_p \mid |x|_p = 1\}$.

4.5 Non-Archimedean Norms as p-Adic Integer Norms in the Unit Ball

4.5.1 Archimedean and Non-Archimedean Absolute Value Properties

An absolute value on a field has the following properties. Following Gouvêa, [90] we write $|\cdot| : \mathbb{K} \to \mathbb{R}^+$, that is, a mapping of the field $\mathbb{K}$ into a non-negative real value. For all $x, y \in \mathbb{K}$ [90, pp. 23–24]:

1. Positive definiteness: $|x| \geq 0$ for all x. $|x| = 0$ if and only if $x = 0$.
2. $|xy| = |x||y|$.
3. $|x + y| \leq |x| + |y|$.

The usual absolute value is such that $|x| = x$ if $x \geq 0$; and $|x| = -x$ if $x < 0$. The foregoing defines the Archimedean absolute value.

Now, though, if the third property is $|x + y| \leq \max\{|x|, |y|\}$, the absolute value is non-Archimedean. A trivial non-Archimedean absolute value is: $|x| = 1$ if $x \neq 0$, and $|0| = 0$.

For p-adic valuations we have these properties:

1. $v_p(xy) = v_p(x) + v_p(y)$.
2. $v_p(x + y) \geq \min\{v_p(x), v_p(y)\}$.

p-adic absolute value is a non-Archimedean absolute value.

4.5.2 A Non-Archimedean Absolute Value, or Norm, is Less Than or Equal to One, and an Archimedean Absolute Value, or Norm, is Unbounded

Equivalent statements are as follows: $\|\cdot\|$ is non-Archimedean; $\|\cdot\| \leq 1$ for every integer n.

This is proved by induction (Katok [107, pp. 11–12]). Firstly, $\|1\| = 1 \leq 1$. Next is the induction step. Take $\|k\| \leq 1$ for all $k \in \{1, 2, \ldots, n-1\}$. We have that $\|n\| = \|(n-1)+1\| \leq$

$\max\{\|n-1\|, 1\} = 1$. Therefore $\|n\| \leq 1$ for all $n \in \mathbb{N}$. Furthermore, since $\| - n\| = \|n\|$, it is concluded that $\|n\| \leq 1$ for all integers $n \in \mathbb{Z}$.

A norm is Archimedean if and only if, given $x, y \in \mathbb{F}$, $x \neq 0$, there is a positive integer, n, such that $\|nx\| > \|y\|$. Exactly the same Archimedean property holds for absolute values. That is, there exists $n \in \mathbb{Z}^+$ such that $|nx| > |y|$.

To see this, with $x, y \in \mathbb{F}$ and $\|y\| > \|x\|$, the Archimedean property implies that $\|n\| > \|y\|/\|x\| > 1$. Conversely, if the norm is Archimedean, there exists a positive integer n where $\|n\| > 1$. We have $\|n\|^k \to \infty$ as $k \to \infty$, and for some k, $\|n^k\| > \|y\|/\|x\|$, which implies the Archimedean property, $\|n^k x\| > \|y\|$ (see [107, p. 12]).

The Archimedean property is equivalent to the following. There are integers with arbitrarily large norms: $\sup\{\|n\| : n \in \mathbb{Z}\} = +\infty$. Again, exactly the same property holds for the Archimedean absolute value. We have that there are integers here that are arbitrarily big.

In the same way, an absolute value (or a norm) is non-Archimedean if and only if $\sup\{|n| : n \in \mathbb{Z}\} = 1$ (Gouvêa [90, p. 31]).

Another result noted by Gouvêa [90, p. 31] is that if $\sup\{|n| : n \in \mathbb{Z}\} = C < +\infty$, then $|\cdot|$ is non-Archimedean and $C = 1$. Another result noted by Katok [107, p. 15] is that when $\|\cdot\|$ is a non-Archimedean norm, then $\|\cdot\|^\alpha$ is also a non-Archimedean norm for any $\alpha > 0$.

4.6 Going Further: Negative *p*-Adic Numbers, and *p*-Adic Fractions

Let

$$a = \sum_{n=0}^{\infty} d_n p^n - (p-1) + (p-1)p + (p-1)p^2 + \ldots + (p-1)p^k + \ldots = \sum_{n=0}^{\infty} (p-1)p^n.$$

Consider $a + 1$, and we find that $a + 1 = 0$. So $a = -1$.

For example, let us subtract 1 from 0 in 7-adics (bearing in mind that the successive digits are, by convention, written leftwards, as here):

```
. . . 0 0 0 0
- . . . 0 0 0 1
---------------
. . . 6 6 6 6
```

Now, with $a = \sum_{n=0}^{\infty}(p-1)p^n$, we can write equally $-1 = (p-1)\sum_{n=0}^{\infty} p^n$. So, from this expression, we have $\sum_{n=0}^{\infty} p^n = \frac{1}{1-p} = \ldots 1111 \in \mathbb{Z}_p$.

Any rational integer is a *p*-adic integer. But there are also (just as in this case) *p*-adic integers among rational fractions. (Note that the *p*-adic expansion of this *p*-adic integer, $\frac{1}{1-p}$, is infinite.)

Next let us look at

$$a = (p-2) + (p-1)p + (p-1)p^2 + \ldots + (p-1)p^k + \ldots.$$

Consider $a + 2$ and we find $a + 2 = 0$. So in this case $a = -2$.

In general, to define $-a$, we take the representation of a, and see what needs to be added to the initial terms of a so that we will have $a + b = 0$, and thereby $b = -a$.

Next topic: $\frac{1}{p}$ has no meaning as a p-adic integer. That is, there is no solution for $p\alpha = 1$. Multiplying by p always gives a p-adic integer ending in 0. (Why? Because the a_0 term necessarily becomes 0. Consider decimal numbers, and multiplying by 10: stick a 0 term at the end. Or for any expansion, we are shifting all terms to the right, when we multiply by p.)

Let p be odd (i.e. $p \neq 2$). Consider

$$\alpha = \frac{p+1}{2} + \frac{p-1}{2}p + \frac{p-1}{2}p^2 + \frac{p-1}{2}p^3 + \ldots + \frac{p-1}{2}p^k + 0.$$

With this consider 2α. Start by considering $\alpha + \alpha$, which we find to be equal to just 1, with all other terms cancelling out and therefore equal to 0. Therefore we find $2\alpha = 1$. So $\alpha = \frac{1}{2}$.

4.7 Number Systems in the Physical and Natural Sciences

In [240], it was noted how "space-time as a manifold with a metric", going back and encompassing the work of Euclid, Riemann and Einstein, has its limitations when faced with lower limits to physical measurement – superstring theory with distances at or below the Planck scale – and gravitational collapse and the cosmological singularity. Now space is a macroscopic notion and requires a large number of particles. To go beyond that, to construct a non-Archimedean geometry (which will be discussed below), the starting point is as follows. Any field contains the field of rational numbers, or the finite Galois field, as a subfield. To develop a physical theory, we require a norm. The rational numbers allow for the usual absolute value as a norm, or the p-adic norm (p prime) which is non-Archimedean (again, this will be discussed further below). There is no non-trivial norm on the finite Galois field. Volovich [240] proposes that number has to be the most basic entity for physical theory, and not particle, quantum field, nor string. When we have $p \to \infty$, then having $\frac{1}{p}$ playing the role of fundamental length allows for tight coupling of the p-adic perspective and the (manifold with a metric) standard theory.

A simple case of the finite Galois field is the integers modulo a prime p, the field $\mathbb{F}_p = \{0, 1, 2, \ldots, p-1\}$. We write $\mathbb{F}_p = \mathbb{Z}/p\mathbb{Z}$, where $\mathbb{Z}$ are the integers. The two simplest fields are therefore the rationals $\mathbb{Q}$ and the finite Galois field, $\mathbb{F}_p$. From these, all that will be discussed below will follow.

It is noted in [240] that it may be unusual to require a physical theory to be constructible over any number field. But the general principle is proposed that "the fundamental physical laws should be invariant under the change of number field". It is suggested [240, p. 85] that "the strongest fluctuations ... in the Big Bang and a newly born Universe can have non-Archimedean or finite or other geometry over non-standard number fields. ... An analogous hypothesis can be considered in the context of the gravitational collapse. By this we mean that in the process of the collapse as a result of quantum effects, matter can collapse into a space with non-Archimedean geometry." Finally, Dirac is quoted as asserting that physics must take the lead from mathematics: "I learnt to distrust all physical concepts as the basis for a theory. Instead one should put one's trust in a mathematical scheme, even if the scheme does not appear at first sight to be connected with physics. The physical meaning had to follow behind the mathematics."

4.8 p-Adic Numbers in Computational Biology and Computer Hardware

Let p denote a prime, and m denote a non-zero positive integer. A p-adic number is such that any set of p integers which are in distinct residue classes modulo p may be used as p-adic digits. Recall that a ring does not, invariably and always, allow division, while a field does. A field is a commutative ring where every non-zero element has an inverse. m-adic numbers form a ring; but p-adic numbers form a field. So *a priori*, 10-adic, called decimal, numbers form a ring. This provides us with a reason for preferring p-adic over m-adic numbers.

We will also be concerned with the ring of integers, $\mathbb{Z}$, and the field of rationals (rational numbers), $\mathbb{Q}$. We have $x \in \mathbb{Q}$ for $x = \frac{a}{b}$, $a, b \in \mathbb{Z}$, $b \neq 0$.

Anther example (which can be considered as analogous to a p-adic or m-adic expansion) is the ring of polynomials with complex coefficients, $\mathbb{C}[X]$, and the field of fractions, $\mathbb{C}(X)$, defined as follows: $f(X) \in \mathbb{C}(X)$, $f(X) = \frac{P(X)}{Q(X)}$, $P(X), Q(X) \in \mathbb{C}[X]$, $Q(X) \neq 0$. So in a sense, $f(X)$ is a "rational function". The p-adic numbers were first developed by Kurt Hensel at the turn of the nineteenth and twentieth centuries. He developed them in close analogy between, on the one hand, the ring of integers, $\mathbb{Z}$, and its field of fractions, the rationals, $\mathbb{Q}$; and on the other hand, the ring of polynomials with complex coefficients, $\mathbb{C}[X]$, and its field of fractions, $\mathbb{C}(X)$. A most readable development of this work is in [90]. (It may be noted that the following has been briefly covered in Section 3.2.4.)

We can consider various p-adic expansions:

1. $\sum_{i=0}^{n} a_i p^i$, which defines positive integers. For a p-adic number, we require $a_i \in 0, 1, \ldots, p-1$.

2. $\sum_{i=k}^{\infty} a_i p^i$, where k is an integer, not necessarily positive, defines the field $\mathbb{Q}_p$ of p-adic numbers.

$\mathbb{Q}_p$, the field of p-adic numbers, is (as seen in these definitions) the field of p-adic expansions.

The choice of p is a practical issue. Indeed, adelic numbers use all possible values of p (see [38] for extensive use and discussion of the adelic number framework). Consider [67, 117]. Deoxyribonucleic acid (DNA) is encoded using four nucleotides: A, adenine; G, guanine; C, cytosine; and T, thymine. In ribonucleic acid (RNA) T is replaced by U, uracil. In [67] a 5-adic encoding is used, since 5 is a prime and thereby offers uniqueness. In [117] a 4-adic encoding is used, and a 2-adic encoding, with the latter based on two-digit boolean expressions for the four nucleotides (00, 01, 10, 11). A default norm is used, based on a longest common prefix – with p-adic digits from the start or left of the sequence.

The relevance of bases other than 10, or 2 for binary arithmetic, can be considered in terms of application domain. Following discussion of number base (and also scale) invariance, Hill [95] briefly comments on octal (base 8) and hexadecimal (base 16), as well as binary, systems for computation. It can also be recalled that a ternary (base 3) computer was developed and built by Nikolay Brusentsov in 1958, and the use of a ternary rather than binary basis for computers remains a subject of interest [106]. The leading mathematician and engineer, Alexey Stakhov, gives detailed and far-reaching discussion of the benefits of ternary number theory for computation (presenting the case for greater robustness of ternary, relative to binary, computers) in [222, 221].

As decimals, the irrational number $\sqrt{3} = 1 + \frac{7}{10} + \ldots$ is an infinite expansion, with decreasing unit of measurement, $m = 10$. The rational number $\frac{1}{3} = \frac{3}{10} + \frac{3}{100} + \ldots$ is also an infinite expansion, with decreasing unit of measurement, $m = 10$.

4.9 Measurement Requires a Norm, Implying Distance and Topology

Whether we deal with Euclidean or with non-Euclidean geometry, we are nearly always dealing with reals. But the reals start with the natural numbers, and from associating observational facts and details with such numbers we begin the process of measurement. From the natural numbers, we proceed to the rationals, allowing fractions to be taken into consideration. (Some of the following comes from Section 3.2.3.)

The following view of how we do science or carry out other quantitative study was proposed by Volovich in 1987 [240, 239] (see also Freund [83]). We can always use rationals to make measurements. The rationals are represented by $\mathbb{Q}$, and we recall that they can be expressed as any integer divided by any other (non-zero) integer. For us, they are "well behaved" and even "tangible" numbers. We have noted already how an infinite decimal series like $0.33333\ldots = 1/3$ is easily expressed as a rational. But there are many numbers that are not, such as π and e (base of Naperian logs), so using rationals alone will allow for an approximation. To go further, we need to allow observables to be "continuous". In practice this means that we can better approximate a non-rational number. The notion of approximation, though, presupposes a topology.

Expressing this slightly differently, observational or experimental measurement will be approximate, in general. It is better therefore to allow for the object of measurement being continuous, which means to endow the objects of measurement with a topology. This implies that we need a completion of the field $\mathbb{Q}$ of rationals. To complete the field $\mathbb{Q}$ of rationals we need Cauchy sequences, and this requires a norm on $\mathbb{Q}$ (because the Cauchy sequence must converge, and a norm is the tool used to show this). There is the Archimedean norm such that, for any $x, y \in \mathbb{Q}$, with $|x| < |y|$, there exists an integer N such that $|Nx| > |y|$. The notation $|\cdot|$ is absolute value. (Informally expressed, it is a measure of "size".) By convention we write $|x|_\infty$ for this norm, the Archimedean norm.

An Archimedean absolute value is a non-negative real value, and satisfies the following axioms: (i) $|x| = 0$ if and only if $x = 0$; (ii) $|xy| = |x||y|$ for all x, y in the field that is considered; (iii) $|x + y| \le |x| + |y|$ for all x, y in the field. Given the absolute value, $|\cdot|$, a distance is $|x - y|$; and $d(x, y)$ is said to be the metric induced by the absolute value.

We next look for alternatives to the Archimedean norm. Remarkably, all norms are known. Besides the Archimedean norm, we have an infinity of norms, $|x|_p$, labelled by primes, p. By Ostrowski's theorem [200] these are all the possible norms on $\mathbb{Q}$. So we have an unambiguous labelling, via p, of the infinite set of non-Archimedean completions of $\mathbb{Q}$ to a field endowed with a topology.

The p-adic norm of a number x makes use of the largest power of p that divides into x. Call that divisor, which is the largest power of p, a valuation. Let us denote it as $v_p(x)$. Then the p-adic norm is $|x|_p = p^{-v_p(x)}$. Extending this to the rationals is done as follows. For $x = a/b$ then the valuation is defined as $v_p(x) = v_p(a) - v_p(b)$. By convention, $v_p(0) = +\infty$. The p-adic norm of x is $|x|_p = p^{-v_p(x)}$.

Ostrowski's theorem tells us that every non-trivial absolute value on $\mathbb{Q}$ is equivalent to one of the absolute values $|\cdot|_p$, where p is a prime number or $p = \infty$.

Here we largely follow [90, pp. 48–49]. Consider x as a positive integer. Consider the p-adic norm for all p including the (Archimedean) ∞ norm. Then

$$\prod_{p \le \infty} |x|_p = 1. \tag{4.1}$$

Since x is a positive integer, we can factor it as $x = p_1^{a_1} p_2^{a_2} \cdots p_k^{a_k}$. So we have $|x|_{p_i} = p_i^{-a_i}$

for $i = 1, 2, \ldots, k$. For any "out of range" $q \neq p_i$, $|x|_q = 1$. (That is, $q > k$, so $|x|_q = q^{-0} = 1/q^0 = 1/1 = 1$.) For the ∞ norm, we have quite simply $|x|_\infty = p_1^{a_1} p_2^{a_2} \cdots p_k^{a_k}$. We see that the overall product provides the result. This result is called the product formula. Beyond positive integers, $\mathbb{Z}^+$ (or beyond non-negative integers, by taking the product formula as yielding 1), the product formula holds for rationals, and it also can be extended to the reals, $\mathbb{R}$, and the complex numbers, $\mathbb{C}$.

4.10 Ultrametric Topology

The ultrametric topology was introduced by Marc Krasner [126] in 1944, the ultrametric inequality having been formulated by Hausdorff in 1934. Essential motivation for the study of this area is provided by [212] as follows. Real and complex fields gave rise to the idea of studying any field K with a complete valuation $|\cdot|$ comparable to the absolute value function. Such fields satisfy the "strong triangle inequality" $|x + y| \leq \max(|x|, |y|)$. Given a valued field, defining a totally ordered Abelian (i.e. commutative) group, an ultrametric space is induced through $|x - y| = d(x, y)$. Various terms are used interchangeably for analysis in and over such fields such as p-adic, ultrametric, non-Archimedean, and isosceles. The natural geometric ordering of metric valuations is on the real line, whereas in the ultrametric case the natural ordering is a hierarchical tree.

A trivial case of a non-Archimedean norm is the case of the trivial absolute value: $|x| = 1$ if $x \neq 0$, and $|0| = 0$.

A number of remarkable properties hold for the ultrametric, also referred to as the ultrametric distance. The ultrametric is a metric, in so far as the strong triangle inequality necessarily implies the metric triangle inequality. Among these remarkable properties are the following, where the proof is briefly indicated (see [137, pp. 41–42]). Note that the following is a summary of what is described in Sections 3.3.1 and 3.3.2.

For any pair of clusters, or balls, these are disjoint or one is a subset of the other.

In the following we can take the ultrametric distance, d, as the absolute value, $|\cdot|$.

In an ultrametric space, every triangle (consisting of three points, x, y, z) is isosceles with small base, or is equilateral. This necessary condition follows from the strong triangle inequality. Take the small base as x and y, with $d(x, z) \leq \max\{d(x, y), d(y, z)\}$. Having $d(x, y) = d(y, z)$ is consistent with this. If we permute x and y, we have $d(y, z) \leq \max\{d(y, x), d(x, z)\}$ and we must have equality in this expression.

Every point of a ball in an ultrametric space is a centre of that ball. To show this, take a as the centre of this ball; take b as an element of the ball; and take x as an element of the ball that is a radius length from b.

All sets, or balls, in an ultrametric space are both open and closed. Topological openness is such that the open ball of radius r and centre a is the set $B(a, r) = \{x : d(x, a) < a\}$. The closed ball, or set, is $\bar{B}(a, r) = \{x : d(x, a) \leq a\}$. A closed set (as defined) contains its boundary. Topologically, too, a closed set is such that its complement is open. A proof of closed sets being open sets, and vice versa, in an ultrametric space, is as follows. We have $B(a, r) \subset \bar{B}(a, r)$. Take a point x, on the boundary of this ball. Look at the open ball, $B(x, s)$ with $s \leq r$. Next look the intersection of the open balls, $y \in B(a, r) \cap B(x, s)$. Due to openness we have $d(y, a) < r$ and $d(y, x) < s \leq r$. Applying the triangle inequality, we have $d(x, a) \leq \max\{d(x, y), d(y, a)\} < \max\{s, r\} \leq r$, so that $x \in \bar{B}(a, r)$ and $x \in B(a, r)$. It follows that balls are both open and closed, written as *clopen*.

For these proofs, just see, for example, [137, pp. 41–43], [90, pp. 35–37] and [107, pp. 53–57].

Finally we note that a topologically zero-dimensional space has a countable base consisting of clopen sets (see [46]). Informally expressed, numbers themselves, or points in an ultrametric space, form no surface or manifold. They are a collection of points. Therefore their dimensionality is 0.

4.11 Short Review of p-Adic Cosmology

First we look at the relevance of both the ∞ norm (or real number system norm) and p-adic norms. All are used.

A fundamental limitation in measurement is when the Planck scale is reached, around 10^{-33} cm. Measurement is granular or non-continuous at or below that scale. In [68], the fundamental time duration, below which measurement of time is not possible, is 10^{-44} seconds.

Whereas real numbers are totally ordered, p-adic numbers have a partial order. The latter comprise a partially ordered set, or poset. Given the ultrametric topology associated with p-adic number systems, we may say then that there is order of inclusion associated with nodes in branches of the tree or hierarchy, but there is no such order (the order of inclusion) between branches. We will return in Section 4.13 below for use of the partial order to provide a scale-free measurement framework.

About 70% of all energy content of the universe consists of dark, unobserved, energy, and about 26% consists of dark, unobserved, mass. Only 4% of our universe is made up of baryons (protons, neutrons). Here we present a short account of the review [66] of possible new ways of accounting for dark energy and dark matter.

In [66] it is proposed that dark energy and dark matter are two forms of p-adic matter. In string theory there are, it is asserted, not only real strings but also p-adic strings. p-adic string theory is discussed in the context of tachyon condensation. In order to use an adelic model, encompassing the reals and all p-adics, a product formula along the lines of equation (4.1) is used. In [66, Section 4.1] product formulas for real and complex numbers are stated. Applications referred to include scattering of two open string tachyons, quantum field theory, string theory, and quantum cosmology. In [66, Section 3.1] it is noted that "ordinary quantum mechanics on a real space can be generalized to quantum mechanics on p-adic spaces for any prime number p".

As noted in previous sections, observation and experiment give measurements that are, first and foremost, elements of the field, $\mathbb{Q}$, of rational numbers. Theoretical models, though, make use of the real numbers, $\mathbb{R}$, that is, the completion of $\mathbb{Q}$ with respect to the distance induced by the ordinary absolute value; and theoretical models make use of the complex numbers, $\mathbb{C}$, that are an algebraic extension of $\mathbb{R}$. So both geometric properties of $\mathbb{Q}$ that are related to possible norms on $\mathbb{Q}$ are used, and algebraic structure of $\mathbb{Q}$.

An adele x is an infinite sequence:

$$x = (x_\infty, x_2, \ldots, x_p, \ldots), \quad x_\infty \in \mathbb{R}, \ x_p \in \mathbb{Q}_p.$$

All but a finite number of these adelic numbers must be $x_p \in \mathbb{Z}_p$ where $\mathbb{Z}_p = \{y \in \mathbb{Q}_p : |y_p| \leq 1\}$ is the ring of p-adic integers. The set of all adeles $\mathbb{A}$ is the following union of direct products:

$$\mathbb{A} = \bigcup_{\mathcal{P}} \mathbb{A}(\mathcal{P}), \quad \mathbb{A}(\mathcal{P}) = \mathbb{R} \times \prod_{p \in \mathcal{P}} \mathbb{Q}_p \times \prod_{p \notin \mathcal{P}} \mathbb{Z}_p.$$

The set of all adeles $\mathbb{A}$ is endowed with componentwise addition and multiplication.

Also defined and discussed in [66, Section 2] is the multiplicative group of ideles, $\mathbb{I}$, where the reals and p-adic numbers considered do not have 0 included.

In [66], we have the section heading "Adelic Universe with Real and p-Adic Worlds" (p. 40). A few pages earlier there is the following motivating statement: "Let us use terms real and p-adic to denote those aspects of the universe which can be naturally described by real and p-adic numbers, respectively. We conjecture here that the visible and dark sides of the universe are real and p-adic ones, respectively" (p. 26).

4.12 Unbounded Increase in Mass or Other Measured Quantity

An unbounded increase in mass, or other measured quantity, is described here as leading to zero or negative outcome value. Just informally expressed, we have saturation. The implications of this, and the insight in this, are fascinating for when we need to consider unbounded or limitless quantities.

In [113, Chapter VI, Section 6] it is sought to use m-adic numbers to describe infinite distance, time interval, and mass, on cosmological scales.

As we have seen in Section 4.6, the p-adic number -1 is

$$-1 = (p-1) + (p-1)p + (p-1)p^2 + (p-1)p^3 + \ldots.$$

To see this, add 1 to each side; then take one term after the other on the right-hand side, and see how term by term this causes the cancelling out of terms. This holds for p-adic and for m-adic numbers.

The previous relation when $p = 2$ is

$$-1 = 1 + 2 + 2^2 + \ldots + 2^n + \ldots.$$

As indicated in [113, p. 99], this implies that if we have "reckoned an infinite number of intervals $\Delta t = 1$ from the initial time moment $t_0 = 0$, we return to the past by one time interval". This implies, referring to m-adic numbers here (which include p-adic numbers), that "in the m-adic model the distinction between the directions in space and that between the past and future disappear".

A further direct consequence of the definition of -1, for $p = 2$, is that $0 = 1 + (1 + 2 + 2^2 + \ldots)$. "Consequently, in the m-adic model being considered the cosmological space–time has the structure of discrete torus of an infinite radius." Due to the two relations that we have looked at, defining -1 and 0, it is considered that: "Infinitely large masses ... are realized in the toroidal m-adic space–time and some configurations with an infinite mass may produce the null effect of a zero ... or negative ... mass" [113, p. 99].

4.13 Scale-Free Partial Order or Hierarchical Systems

In Altaisky and Sidharth [3] it is noted how the real numbers are "a theoretical abstraction valid only at large classical scales" (i.e. above the Planck scale). Mention is made of space–time at small scales being granular. We may note too that, going back in time, a minimum irreducible unit of time, the chronon, had been proposed [3].

The authors in [3] proceed to describe a p-adic quantum field theory. Without background space, and if there is only a static view taken, then there is no time either. Geometry requires multiple objects. Hence the viewpoint that reaching back to the Big Bang, and the standard cosmological model of the early Universe, are an extrapolation of today's structure of the Universe. Instead relations and interrelationships between objects are to be considered as members of sets, and in the "descendent" relation between sets.

How do we have a non-Archimedean geometry at Planck scales, since space-time is granular there? It is shown in [3] how points on the sphere can be put in one-to-one correspondence with the space of p-adic integers, $\mathbb{Z}_p$. This is using an infinite sequence of embedded triangles scheme; the triangle itself, when glued along its edges, furnishes the sphere. The two-dimensional sphere, S^2, is at issue here. This leads then to a p-adic loop integral.

What is used here is a triangulation of a spherical surface, in order to provide a coordinate system that is hierarchical (in terms of the partially ordered or tree-structured triangulation). In fact this work is reminiscent of what is done, algorithmically, to perform interpolation of data on the surface of a sphere.

More generally, any point in a physical space can be located by the nested set of its vicinities. The spaces of the nested family do not form a σ-algebra (as used in probability theory) but their differences do; as a consequence, integration is defined on the sum of differences.

The implication is that, on the one hand, description of the Universe is in terms of locally Euclidean coordinates. On the other hand, description of the Universe is in terms of the nested set of vicinities. This nested set is such that each vicinity (ball) is contained in another, or is of empty intersection. Partial intersections are not allowed.

This suggests for galaxy-scale structures that ancestor distance is also important compared to distance measured by travelling light waves. The set of vicinities is countable; and union of vicinities is a complete Hausdorff space; these imply that union of vicinities is metrizable due to the Urysohn lemma. (The Urysohn function f on a topological space, X, with any two non-empty, closed, disjoint subsets, A, B, is a continuous function, $f : X \to [0, 1]$ with $f(A) = \{0\}$, $f(B) = \{1\}$.) Time, or time-like distance, is distance measured with respect to branching points. There is derivation (up to a constant) of "mass = inverse of probability" (i.e. inverse probability amplitude of transition between objects). This approach is coordinate-free in that there are no assumptions about space before or beyond matter.

For large-scale structures in the Universe, it is concluded that: "The distance between different objects, even between galaxies, may therefore be measured not only by travelling light waves, but also by the level of their common ancestor in the evolution process." The "quantum time point of view" measures time by number of branching points. Inertial mass in quantum mechanics is then looked at on a discrete set. The discussion in [3] is very much therefore based in p-adic expression of an ultrametric (or hierarchical) topology.

It is concluded as follows, that this approach "is coordinate-free and thus no assumption about the existence of any space before or beyond matter were used". Once p-adic expression of object relationships is feasible, "we can introduce the mass as the inverse probability amplitude of transition between these objects, and further the concept of metric also arises naturally". Mass and metric are derived from quantum mechanics in this p-adic (and ultrametric) context.

4.14   *p*-Adic Indexing of the Sphere

For cosmology, it has been noted above how Altaisky and Sidharth [3] have proposed a relative coordinate system, rather than an absolute frame of reference. A consequence of relative or local measurement is that neighbourhood relations become scale-invariant in the following sense. Large zones (in the manner in which they are specified) can be neighbours, and less large zones can be neighbours. All depends therefore on the zone, or region, or node in the tree.

In [229], a hierarchical mesh encoding of spherical astronomical or cosmological data is pursued. The authors start with an octahedral representation of a sphere. This is, for the northern hemisphere, four triangles, and for the southern hemisphere, four triangles. One could just as easily use a six-sided figure, with three triangles in the northern hemisphere, and three triangles in the southern hemisphere. Irrespective of the eight- or six-fold set of triangles, let us look at the refinement of these triangles. These are to be adaptive (i.e. tangential) to the subjacent (underlying) sphere.

Figure 4.1 illustrates the hierarchical (that is to say, partially ordered set) subdivision, which is also a partitioning. It is to be noted how a *p*-adic ($p = 5$) set of digits describe a path from root to terminal (i.e. zone) of specified level of precision. Figure 4.1 shows the first and second levels. Labels for the nodes in the regular five-way tree are shown.

In [229] the encoding starts with the octahedral representation of the sphere (the cosmos, and that could well be also the planetary globe). Therefore there are eight planes that are tangential to the underlying sphere. Then each of those triangular, tangential planes is recursively refined, as is illustrated in Figure 4.1. The number of triangle nodes is then given as, for depth d, $8 \times 4^{d-1}$. Many querying algorithms are then discussed (determining convex regions, ranges, intersections, and so on).

4.15   Diffusion and Other Dynamic Processes in Ultrametric Spaces

In [197] random walks in ultrametric spaces are studied. An ultrametric distance matrix can be used to define transition probabilities. Thus there are "locally stable states separated by energy barriers defining an ultrametric topology". The ultrametric space, with a hierarchical tree representing it, consists of energy basins, that can be embedded. So these energy basins are locally stable states. Then the "transitions between states are thermally activated with rates determined by the ... energy barriers separating the states." The probability of a particle being at site i at time t is given by $p_i(t)$. In a hierarchical tree with n terminal nodes, and $n-1$ non-terminal nodes (let us assume, formed by a hierarchical agglomerative clustering of the state's characteristics), there are $2n-1$ states.

In [11] a protein energy landscape is considered, with local minima corresponding to the protein conformational substates. In [154] the description is of the dynamics of protein folding, through conformational fluctuation. General, hierarchical energy landscapes are described in [12], together with *p*-adic analysis of diffusion.

Further discussion can be found in [4]. Areas covered in relation to dynamics in *p*-adic spaces include cognitive science and neuroscience, cryptography and also computational biology. Dynamical systems are studied in all of these diverse fields.

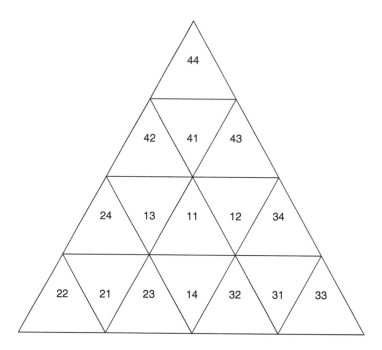

FIGURE 4.1: Example of a *p*-adic encoding of a hierarchical triangular mesh or tree covering. At the first level, the triangle zones have labels 1, 2, 3, 4. At the next level, the triangle with label 2 is partitioned into triangles with labels 21, 22, 23, 24.

Part III

New Challenges and New Solutions for Information Search and Discovery

5

Fast, Linear Time, m-Adic Hierarchical Clustering

5.1 Pervasive Ultrametricity: Computational Consequences

Both hierarchy and geometry (e.g. Euclidean factor space arising out of correspondence analysis) support the understanding of, and working in, complex systems. Applications are overviewed for search and discovery, information retrieval, clusterwise regression, and knowledge discovery.

5.1.1 Ultrametrics in Data Analytics

A brief statement of basic ideas and definitions is as follows. Metric geometric and ultrametric topology have their complementary roles. (1) Euclidean geometry expresses the semantics of information. Because Euclidean, it is straightforward to visualize results and outcomes. (2) Hierarchical topology expresses other aspects of semantics, and in particular how a hierarchy expresses anomaly or change. A further useful case is when the hierarchy respects chronological or other sequence information.

One enabler, useful among other spectral analysis (or dimensionality reduction) techniques, and offering an extensive geometry- and topology-based methodological framework and platform, is correspondence analysis. This can be taken as a tale of three metrics [171]. There is a long tradition of research in metric methods for data analysis. The ultrametric perspective too has a long tradition, having come into data analysis in the 1960s.

Correspondence analysis is a latent semantic analysis [216]. Typically in latent semantic analysis or latent semantic indexing, the matrix to be diagonalized is weighted by term frequency and inverse document frequency terms. In correspondence analysis the analogous weighting is carried out using the row and column masses, that is, the densities of the observation set and the attribute set, respectively. The effect of this in correspondence analysis is that the clouds of observations and of attributes are endowed with the χ^2 distance. We have $d^2(i, k) = \sum_j \frac{1}{f_j}(f_{ij}/f_i - f_{kj}/f_k)^2$ for observations i, k; attributes j; f_{ij} the original non-negative data values (e.g. frequencies of occurrence, or presence/absence values) that have been normalized by dividing by the grand total, $\sum_i \sum_j f_{ij}$, and finally the row masses, $f_i = \sum_j f_{ij}$, and analogously the column masses, f_j. Eigenreduction follows to find the principal axes of inertia, yielding the factors. We consider, for the two clouds of points in the dual spaces (of observations and of attributes, sharing the same eigenvalues) the projections, the relative contribution to inertia (by point and by factor), and the relative correlation (by point and by factor).

A small additional note is in order. Correspondence analysis is applicable to non-negative data, but more generally it is applicable to cross-tabulated data where the marginal sums are non-negative (see [131, pp. 59–60]).

Thus we have the following: (1) the χ^2 metric, appropriate for profiles of frequencies of occurrence, or presence/absence data, among other types of data; (2) the Euclidean metric, for visualization, and for static context; (3) the ultrametric, for hierarchic relations and for dynamic, in the sense of chronologically ordered, context. The triangle inequality holds for metrics: $d(x, z) \leq d(x, y) + d(y, z)$.

Euclidean distance makes a lot of sense when the population is homogeneous, especially for allowing planar visualizations. Ultrametric distance makes a lot of sense when the observables are heterogeneous, or discontinuous. Considering the latter, data endowed with the ultrametric distance, is especially useful for determining: anomalous, atypical, innovative cases. There follows here, a short summary of all that is described in Section 3.3.1 and following subsections. The strong triangle inequality, or ultrametric inequality, holds for such tree distances: $d(x, z) \leq \max\{d(x, y), d(y, z)\}$. An example of an ultrametric is a closest common ancestor distance. Some properties of an ultrametric space follow: (i) all triangles are isosceles with small base, or equilateral; (ii) every point in a ball is its centre; (iii) the radius of a ball equals the diameter; (iv) a ball is *clopen* (viz. topologically both closed and open); (v) an ultrametric space is always topologically 0-dimensional. Informally expressed, (iv) is due to both being with the boundary, and being outside the boundary of any neighbouring ball; and (v) is due to points and all their relationships alone being at issue and not lines, surfaces and hypersurfaces.

In regard to a ball being *clopen* in an ultrametric topology, Levin gives the following informal summary [138]: "Approximation has two subtly different aspects: metric and topology. Metric tells how close our ideal point is to a specific wrong one. Topology tells how close it is to the combination of all unacceptable (non-neighboring) points."

5.1.2 Quantifying Ultrametricity

Summary examples of application to text collections and time series follow. Further elaboration and discussion are in Chapter 9.

1. Take all triplets of points (or sample all triplets), check isosceles with small base, or equilateral, configurations, and determine a coefficient of the relative proportion of such triangles. We refer above to isosceles with small base, or equilateral, triangles in a Euclidean perspective as a property of an ultrametric space.

2. On a scale from $1 = 100\%$ ultrametric-respecting properties to $0 = 0\%$, we found the following averaged values for works of literature and other texts: Grimm Brothers, 0.1147; Jane Austen, 0.1404; aviation accident reports, 0.1154; dream reports, 0.1933 (in the case of one person, 0.2603); the value for Joyce's *Ulysses* lay between the latter two. See [180].

3. For one-dimensionaal signals the following results were found [172]: FTSE, USD/EUR, sunspot, stock, futures, eyegaze, Mississippi, www traffic, EEG/sleep/normal, EEG/petit mal epilepsy, EEG/irregular epilepsy, quadratic chaotic map, uniform. Eyegaze data was found to be high, and the chaotic series was found to be low, in inherent ultrametricity.

5.1.3 Pervasive Ultrametricity

1. As dimensionality increases, so does ultrametricity.

2. In very high-dimensional spaces, the ultrametricity approaches 100%.

3. Relative density is important: high-dimensional and spatially sparse mean the same in this context.

For discussion and simulations, see [92, 169].

5.1.4 Computational Implications

Consider a dendrogram: a rooted, labelled, ranked, binary tree. So we have n terminals and $n-1$ levels.

A dendrogram's root-to-terminal path length is $\log_2 n$ for a balanced tree, and $n-1$ for an imbalanced tree. Call the computational cost of such a traversal $O(t)$, where t is this path length. The inequality $1 \geq O(t) \geq n-1$ holds.

Adding a new terminal to a dendrogram is carried out in $O(t)$ time.

The cost of finding the ultrametric distance between two terminal nodes is twice the length of a traversal from root to terminals in the dendrogram. Therefore distance is computed in $O(t)$ time.

Nearest neighbour (or best match) search in ultrametric space can be carried out in $O(1)$ or constant time. This is due to either (i) the closest pair sharing a common parent node, or (ii), given that the tree is binary, there being a common parent node for the given terminal, and a pair of terminals.

5.2 Applications in Search and Discovery using the Baire Metric

The approach to hierarchical clustering to be described is neither agglomerative nor divisive (both of which terms have long been used as a summary of major algorithms), but rather a matter of *hierarchical encoding* of data.

We review in this section the inducing of an ultrametric topology through use of the Baire distance, for clustering of large data sets. This includes hierarchical clustering via Baire distance using Sloan Digital Sky Survey (SDSS) spectroscopic data, and hierarchical clustering via Baire distance using chemical compounds.

5.2.1 Baire Metric

The Baire metric, or longest common prefix distance, is simultaneously an ultrametric. Consider a number as a sequence of digits, for example the pair of numbers $\{x_k, y_k \mid k = 1, 2, 3, 4\}$, and define for base b their Baire distance from $\kappa = \mathrm{argmax}_k\{x_k = y_k\}$ and $d_b = b^{-\kappa}$. For example, for $x = 0.4254$, $y = 0.4257$, $b = 10$, we have $d_{10} = 10^{-3}$.

The Baire (ultra)metric space consists of countable infinite sequences with a metric defined in terms of the longest common prefix [13]. The longer the common prefix, the closer a pair of sequences.

Given that the Baire distance is an ultrametric distance, it follows that a hierarchy can be used to represent the relationships associated with it. Furthermore, the hierarchy can be directly read from a linear scan of the data. It can be termed a hierarchical hashing scheme.

5.2.2 Large Numbers of Observables

The following astronomy data included right ascension and declination coordinates (the extra-solar system latitude and longitude) and redshifts (cosmological distance, time lookback, and recession velocity). For the latter, both the less resource-demanding photometric redshift and the more resource-demanding spectrometric redshift were available.

From the SDSS data, we took a subset of approximately 0.5 million data points from SDSS release 5 [6]. From this a selection was made of declination and right ascension coordinates, together with (higher-precision) spectrometric redshift, z_{spect}, and (lower-precision, more easily measured) photometric redshift, z_{phot}. An example of these values, respectively right ascension coordinate, declination coordinate, spectrometric redshift and photometric redshift, was $145.4339, 0.56416792, 0.14611299, 0.15175095$. The aim in this case study was to regress z_{spect} on z_{phot}.

Towards this aim, we further sought to determine good-quality mappings of z_{spect} onto z_{phot}, that is, a form of clusterwise nearest neighbour regression. This is not spatially (in right ascension and declination) clusterwise but rather within the data set itself.

The framework for fast clusterwise regression led to the following outcome:

- 82.8% of z_{spect} and z_{phot} have at least 2 common prefix digits (i.e. numbers of observations sharing 6, 5, 4, 3, 2 decimal digits). We can find very efficiently where these 82.8% of the astronomical objects are.

- 21.7% of z_{spect} and z_{phot} have at least 3 common prefix digits (i.e. numbers of observations sharing 6, 5, 4, 3 decimal digits).

This SDSS study is described in more detail in [51]. See also the book's website for this data set, and for analysis software.

5.2.3   High-Dimensional Data

We turn now to another case study, using chemoinformatics, which is high-dimensional. Since we are using digits of precision in our data (re)coding, we must address how we handle high dimensions. This work is described in detail in [186].

We have developed an approach for the Baire distance applied to chemical compounds. The matching of chemical structures has the following aims. Clustering of compounds based on chemical descriptors or chemical representations, in the pharmaceutical industry, are used for screening large corporate databases. Chemical warehouses are expanding due to mergers, acquisitions, and the synthetic explosion brought about by combinatorial chemistry.

Binary (boolean presence/absence-valued) "fingerprints" are defined as fixed-length bit strings for a given chemical compound. We used 1.2 million chemical compounds, each characterized by 1052 boolean presence/absence values.

We normalize chemical compounds by dividing each row by its row sum (hence giving the "profiles" of chemical compounds, in correspondence analysis terms).

5.2.4   First Approach Based on Reduced Precision of Measurement

In this approach we use the fact that precision of measurement leads to greater ultrametricity (i.e. the data are more hierarchical). From this we develop an algorithm for finding equivalence classes of specified precision chemicals.

We thus limit thq precision of compound/attribute values. This has the effect of more chemical compound values becoming the same for a given attribute. Through a heuristic (e.g. interval of row sum values), equivalence classes of zero-distance compounds can be read off, with restricted precision.

Further analysis of these crude clusters can be done if required. We call this "data condensation". A range of evaluations were carried out on subsets of 20,000 compounds and 1052 attributes.

5.2.5 Random Projections in High-Dimensional Spaces, Followed by the Baire Distance

In this second approach for processing our chemoinformatics data, we use random projections of the 1052-dimensional space in order to find the Baire hierarchy. We find that clusters derived from this hierarchy are quite similar to k-means clustering outcomes.

In fact random projection here works as a class of hashing function. Hashing is much faster than alternative methods because it avoids the pairwise comparisons required for partitioning and classification. If two points (p, q) are close, they will have a very small $|p - q|$ (Euclidean metric) value; and they will hash to the same value with high probability; if they are distant, they should collide with small probability.

5.2.6 Summary Comments on Search and Discovery

As outlined, we have a new way of inducing a hierarchy on data. The first viewpoint is that we encode the data hierarchically and essentially read off the clusters.

An alternative viewpoint is that we can cluster information based on the longest common prefix.

We obtain a hierarchy that can be visualized as a tree. We are hashing, in a hierarchical or multiscale way, our data.

We are targeting clustering in massive data sets. The Baire method offers a fast alternative to k-means and *a fortiori* to traditional agglomerative hierarchical clustering. At issue throughout this work is the embedding of our data in an ultrametric topology.

In some respects, ultrametric topology simply expresses, in a coherent and helpful way, what is understood also in other ways (cf. how the Baire hierarchy can be viewed as a hashing or binning of our data). Nonetheless our examples and case studies serve to show how powerful ultrametric topological analytics are. Furthermore, such ultrametric analytics are highly scalable. Note again that data and software are available on the book's website.

5.3 m-Adic Hierarchy and Construction

Here, m-adic hierarchical clustering is for applications that include search and retrieval using the Baire metric, with linkages to many related topics such as hashing and precision properties of data measurement. Various vantage points on the Baire metric are described, and its use in clustering data, or its use in preprocessing and structuring data in order to support search and retrieval operations. Often we can proceed directly to clusters, with no need to determine the distances. It is shown how a hierarchical clustering can be read directly from one pass through the data. We offer insights also on practical implications of precision of data measurement. As a mechanism for treating multidimensional data, including very high-dimensional data, random projections are used.

In areas such as search, matching, retrieval and general data analysis, massive increase in data requires new methods that can cope well with the explosion in volume and dimensionality. In this work, the Baire metric, which is furthermore an ultrametric, is used to induce a hierarchy and in turn to support clustering, matching and other operations.

Arising directly out of the Baire distance is an ultrametric tree, which also can be seen as a tree that hierarchically clusters data. This presents a number of advantages when storing and retrieving data. When the data source is in numerical form this ultrametric tree can be used as an index structure making matching and search, and thus retrieval, much easier.

The clusters can be associated with hash keys, that is to say, the cluster members can be mapped onto *bins* or *buckets*.

Another vantage point in this approach is precision of measurement. Data measurement precision can be either used as given or modified in order to enhance the inherent ultrametric and hence hierarchical properties of the data.

Rather than mapping pairwise relationships onto the reals, as distance does, we can alternatively map onto subsets of the power set of, say, attributes of our observation set. This is expressed by the generalized ultrametric, which maps pairwise relationships into a partially ordered set. It is also current practice as formal concept analysis where the range of the mapping is a lattice (see Section 5.9).

Relative to other algorithms the Baire-based hierarchical clustering method is fast. It is a direct reading algorithm involving one scan of the input data set, and is of linear computational complexity.

Many vantage points are possible, all in the Baire metric framework. Many of the following vantage points have been noted and observed in earlier chapters.

- A metric that is simultaneously an ultrametric.

- Hierarchy induced through m-adic encoding (m a positive integer, e.g. 10).

- p-adic (p prime) or m-adic clustering.

- Hashing of data into bins.

- Data precision of measurement implies how hierarchical the data set is.

- Generalized ultrametric.

- Lattice-based formal concept analysis.

- Linear computational time hierarchical clustering.

5.4 The Baire Metric, the Baire Ultrametric

5.4.1 Metric and Ultrametric Spaces

Essential properties are covered in other chapters and summarized here. Quite central to analytics, with a topological perspective, is the mapping of data into an ultrametric space or, alternatively expressed, searching for an ultrametric embedding, or ultrametrization [237]. Actually, inherent ultrametricity leads to an identical result relative to most commonly used agglomerative criteria [167]. Furthermore, data coding can help greatly in finding how inherently ultrametric data are [169], and this is further discussed in Section 5.8.

A metric space (X, d) consists of a set X on which is defined a *distance function* d which assigns to each pair of points of X a distance between them, and satisfies the following four axioms for any triplet of points x, y, z:

1. $\forall\, x, y \in X,\ d(x, y) \geq 0$ (positiveness).

2. $\forall\, x, y \in X,\ d(x, y) = 0$ if and only if $x = y$ (reflexivity).

3. $\forall\, x, y \in X,\ d(x, y) = d(y, x)$ (symmetry).

4. $\forall\, x, y, z \in X,\ d(x, z) \leq d(x, y) + d(y, z)$ (triangle inequality).

An ultrametric space respects the strong triangle inequality, or ultrametric inequality, defined as

$$d(x, z) \leq \max\{d(x,y),\ d(y,z)\},$$

for any triplet of points $x, y, z \in X$ in addition to the positive definiteness properties of pairs of points, expressed above.

Various properties of an ultrametric space ensue from this. For example, the triangle formed by any triplet is necessarily isosceles, with the two large sides equal, or is equilateral. Every point of a circle in an ultrametric space is a centre of the circle. Two circles of the same radius, that are not disjoint, are overlapping [4, 137]. Additionally, an ultrametric is a distance that is defined strictly on a tree, which is a property that is very useful in classification [25].

5.4.2   Ultrametric Baire Space and Distance

A Baire space consists of countably infinite sequences with a metric defined in terms of the longest common prefix: the longer the common prefix, the closer a pair of sequences. What is of interest to us is this longest common prefix metric, which is termed the Baire distance [35, 158, 186].

Begin with the longest common prefixes at issue being digits of precision of univariate or scalar values. For example, let us consider two such decimal values, x and y, with both measured to some maximum precision. We take as maximum precision the length of the value with the fewer decimal digits. With no loss of generality we take x and y to be bounded by 0 and 1. Thus we consider ordered sets x_k and y_k for $k \in K$. So $k = 1$ is the first decimal place of precision, $k = 2$ is the second decimal place, and so on; $k = |K|$ is the $|K|$th decimal place. The cardinality of the set K is the precision with which a number, x or y, is measured.

Consider as examples $x_3 = 0.478$ and $y_3 = 0.472$. Start from the first decimal position. For $k = 1$, we find $x_1 = y_1 = 4$. For $k = 2$, $x_2 = y_2 = 7$. But for $k = 3$, $x_3 \neq y_3$.

We now introduce the following distance (in the case of vectors x and y with one attribute, hence unidimensional):

$$d_{\mathcal{B}}(x_K, y_K) = \begin{cases} 1 & \text{if } x_1 \neq y_1, \\ \inf\ \mathcal{B}^{-n} & \text{if } x_n = y_n,\ \ 1 \leq n \leq |K| \end{cases} \tag{5.1}$$

We call this $d_{\mathcal{B}}$ value Baire distance, which is a 1-bounded ultrametric [35, 186] distance, $0 < d_{\mathcal{B}} \leq 1$. When dealing with binary (boolean) data 2 is the chosen base, $\mathcal{B} = 2$. When working with real numbers the base is best defined to be 10, $\mathcal{B} = 10$. With $\mathcal{B} = 10$, for instance, it can be seen that the Baire distance is embedded in a 10-way tree which leads to a convenient data structure to support search and other operations when we have decimal data. As a consequence data can be organized, stored and accessed very efficiently and effectively in such a tree.

For $\mathcal{B}$ prime, this distance has been studied by Benois-Pineau et al. [23] and Bradley [37, 35], with many further (topological and number-theoretic, leading to algorithmic and computational) insights arising from the p-adic (where p is prime) framework.

5.5 Multidimensional Use of the Baire Metric through Random Projections

It is well known that traditional clustering methods do not scale well in very high-dimensional spaces. A standard and widely used approach when dealing with high dimensionality is to apply a dimensionality reduction technique. This consists of finding a mapping F relating the input data from the space $\mathbb{R}^d$ to a lower-dimension feature space $\mathbb{R}^k$: hence, $F : \mathbb{R}^d \to \mathbb{R}^k$.

A least-squares optimal way of reducing dimensionality is to project the data onto a lower-dimensional orthogonal subspace. Principal component analysis (PCA) is a popular choice to do this. It uses a linear transformation to form a simplified (i.e. reduced-dimensionality) data set while retaining the characteristics (i.e. variances) of the original data. PCA selects a best-fitting, ordered sequence of subspaces (of dimensionality $1, 2, 3, \ldots$) that best preserve the variance of the data.

This is a good solution when the data allow these calculations, but PCA as well as other dimensionality reduction techniques remain expensive from a computational point of view for very large data sets. The essential eigenvalue and eigenvector decomposition is of $O(d^3)$ computational complexity. Looking beyond the least-squares and orthogonal PCA projection, we have studied the benefits of random projections.

Random projection [32, 54, 57, 75, 81, 141, 144, 238] is the finding of a low-dimensional embedding of a point set, such that the distortion of any pair of points is bounded by a function of the lower dimensionality.

The theoretical support for random projection can be found in the Johnson–Lindenstrauss lemma [105]. It states that a set of points in a high-dimensional Euclidean space can be projected into a low-dimensional Euclidean space such that the distance between any two points changes by a fraction of $1 + \varepsilon$, where $\varepsilon \in (0, 1)$.

Johnson–Lindenstrauss Lemma. For any $0 < \varepsilon < 1$ and any integer n, let k be a positive integer such that

$$k \geq 4(\varepsilon^2/2 - \varepsilon^3/3)^{-1} \ln n. \tag{5.2}$$

Then for any set V of any points in $\mathbb{R}^d$, there is a map $f : \mathbb{R}^d \to \mathbb{R}^k$ such that for all u, $v \in V$,

$$(1 - \varepsilon) \parallel u - v \parallel^2 \ \leq \ \parallel f(u) - f(v) \parallel^2 \ \leq \ (1 + \varepsilon) \parallel u - v \parallel^2 .$$

Furthermore, this map can be found in randomized polynomial time [55].

The original proof [105] was further simplified by Frankl and Maehara [82] and Dasgupta and Gupta [55] (see also [1, 238]).

In practice we find that random directions of high-dimensional vectors are a sufficiently good approximation to an orthogonal system of axes. In this way we can exploit data sparsity in high-dimensional spaces.

In random projection the original d-dimensional data are projected to a k-dimensional subspace ($k \ll d$), using a random $k \times d$ matrix R:

$$X'_{k \times N} = R_{k \times d} \, X_{d \times N}, \tag{5.3}$$

where $X_{d \times N}$ is the original set with d-dimensionality and N observations. From the computational aspect, forming the random matrix R and projecting the $d \times N$ data matrix X into the k dimensions is of order $O(dkN)$. If X is sparse with c non-zero entries per column, the complexity is of order $O(ckN)$.

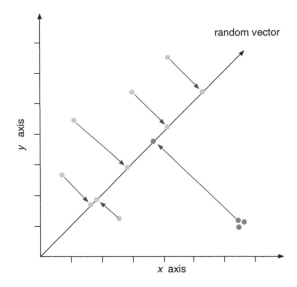

FIGURE 5.1: Random projection axis, showing orthogonal projections.

Random projection can be seen as a class of hashing function (this is further discussed in Section 5.7 below). Hashing is much faster than alternative methods because it avoids the pairwise comparisons required for partitioning and classification. This process is depicted in a Euclidean two-dimensional space in Figure 5.1, where a random vector is drawn and data points are projected onto it. If two points (p, q) are close, they will have a very small $\|p-q\|$ (Euclidean metric) value; and they will hash to the same value with high probability. If they are distant, they should collide with small probability.

5.6 Hierarchical Tree Defined from m-Adic Encoding

Consider data values, base 10, that are greater than 0 and less than 1. Let the full data set be associated with the root of a regular 10-way tree. Determine 10 clusters/bins at the first level from the root of the tree, labelled through the first digit of precision, $0, 1, 2, \ldots, 9$. Determine the first level of the tree – for each of the 10 first-level clusters – labelled through the second digit of precision. The clusters/bins, associated with terminals in the tree, can be labelled $00, 01, 02, \ldots, 09; 10, 11, \ldots, 19; 20, 21, \ldots, 29; \ldots, 90, 91, 92, \ldots, 99$. This m-adic encoding tree, with $m = 10$, can be continued through further levels.

In [51], a large cosmology survey data set is used, where it is sought to match spectrometric redshifts against photometric redshifts. These are respectively denoted z_{spec} and z_{phot}. For convenience, some essential aspects of the earlier description in this chapter are repeated in the following.

The commonly used, extra-solar system, coordinate frame uses three spatial dimensions – right ascension and declination, and redshift. Redshift is therefore one dimension of the three-dimensional cosmos. Apart from cosmological distance, it can also be viewed in terms of recession velocity, and look-back time to the observation in question. Using spectrometry is demanding in terms of instrumental procedure, but furnishes better-quality output (in terms of precision and error). Photometric redshifts, on the other hand, are more easily obtained but are of lower quality. Hence there is interest in inferring spectrometric redshifts from photometric redshifts. With that aim in mind, in [51] we looked at a clusterwise regression approach, where "clusterwise" is not spatial but rather based on measurement precision.

To summarize our results on approximately 400,000 pairs of redshift measurements, we found the following.

- 82.8% of z_{spec} and z_{phot} have at least 2 common prefix digits. This relates to numbers of redshift couples sharing the first digit, and then the first and second decimal digits. We can find very efficiently where these 82.8% of the astronomical objects are in our data:

- 21.7% of z_{spec} and z_{phot} have at least 3 common prefix digits. This relates to numbers of redshift couples sharing the first digit, and then the first, second and third decimal digits.

This exemplifies how we read off clusters from the hierarchical tree furnished by the Baire (ultra)metric.

5.7 Longest Common Prefix and Hashing

The longest common prefix, used in the Baire metric, can be viewed as a hashing or data binning scheme. We will follow up first on the implications of this when used in tandem with random projections for multivariate data (i.e. data in a Euclidean or other space of dimensionality greater than 1).

5.7.1 From Random Projection to Hashing

Random projection is the finding of a low-dimensional embedding of a point set – dimension equals 1, hence a line or axis, in this work – such that the distortion of any pair of points is bounded by a function of the lower dimensionality [238]. As noted in Section 5.5, there is an extensive literature in this area (e.g. [70]). While random projection *per se* will not guarantee a bijection of best match in original and in lower-dimensional spaces, our use of projection here is effectively a hashing method. We aim to deliberately find hash collisions that thereby provide a sufficient condition for the mapped vectors to be matched. Alternatively expressed, candidate best match vectors are determined in this way. As an example of this approach, Miller et al. [155] use the MD5 (Message Digest 5) hashing scheme as a basis for nearest neighbour searching. Buaba et al. [41] note that hashing is an approximate matching approach whereby (i) the probability of not finding a nearest neighbour is very small, and (ii) neighbours in the same hash class furnished by the hashing projection are good approximations to the (optimal) nearest neighbour. Retrieval of audio data is at issue in [155], retrieval of Earth observation data is studied in [41], and content-based image retrieval is the focus of [246].

A hash function, used for similarity finding, maps data into a fixed-length data *key*. Thus, possibly following random projection, assume our data set includes two strings of values 0.457891 and 0.457883456. Now consider how both of these can be put into the same "bin" or "bucket" labelled by 4578. In this case the fixed-length hash key of a data value is read off as the first 4 significant digits.

Collision of identically valued vectors is guaranteed, but what of collision of non-identically valued vectors, which we want to avoid? Such a result can be established based on the assumption of what distribution our original data follow. A stable distribution is used in [98], that is, a distribution such that a limited number of weighted sums of the variables is also itself of the same distribution. Examples include both Gaussian (which is 2-stable, [97]) and power law (long-tailed) distributions.

Interestingly, however, very high-dimensional (or equivalently, very low sample size or low N) data sets, by virtue of high relative dimensionality alone, have points mostly lying at the vertices of a regular simplex or polygon [92, 169]. Such regular sparsity is one reason why we have found random projection to work well. Another reason is that we use attribute weighting (e.g. [186]). Li et al. [141, Section 5] note that pairwise distance can be "meaningless" when using heavy-tailed distributions, but this problem is bypassed by attribute weighting which modifies second- and higher-order moments of the data. Thereafter, with the random projection mapping using statistically uniformly drawn weights for attribute j, w_j, the random projections for data vectors x and y are respectively $\sum_j w_j x_j$ and $\sum_j w_j x'_j$. We can anticipate near equal x_j and x'_j terms, for all j, to be mapped onto fairly close resultant scalar values.

We adopted an experimental approach to confirm these hypotheses, namely, that high-dimensional data are "regular" or "structured" in such a way; and that, as a consequence, hashing is particularly well behaved in the sense of non-identical vectors being nearly always collision-free. We studied stability of results, and also effectiveness relative to other clustering methods, in particular k-means partitioning. In [186], this principle of binning data is used on a large, high-dimensional chemoinformatics data set. It is shown in [51] how a large astronomical data set also lends itself very well to similarity finding in this way.

5.8 Enhancing Ultrametricity through Precision of Measurement

By reducing the precision of measurement we are in effect mapping the data into bins, or hashing the data. Furthermore, the longest common prefix as used in the Baire metric gives us one way, that is both practical and useful, to extract reduced-precision information from our data.

5.8.1 Quantifying Ultrametricity

We assume Euclidean space. If necessary our data can be mapped into a Euclidean space, which maps contingency table data endowed with the chi-squared metric into a Euclidean factor space. In a different application domain where data are given as pairwise comparisons or preferences, multidimensional scaling methods take pairwise ranks (hence ordinal data) and perform a mapping into a Euclidean space.

Consider, in Euclidean space, a triplet of points that defines a triangle. Take the smallest internal angle in the triangle, $a \leq 60$ degrees. For the two other internal angles, b and c, if $|b - c| < 2$ degrees then we characterize the triangle as being approximately isosceles with small base, or equilateral. That is to say, we consider 2 degrees to be an arbitrary small angle. Because of the approximation involved we could claim, informally, that this leads to a fuzzy definition of ultrametricity.

Any triangle in Euclidean space is ultrametric if it is isosceles with small base, or equilateral [137].

To use this in practice, we look for the overall proportion of such triangles in our data. This yields a coefficient of ultrametricity [169]. Therefore to quantify ultrametricity we take all possible triplets, i, j, k. We look at their angles, and judge whether or not the ultrametric triangle properties are verified. Having examined all possible triangles, our ultrametricity

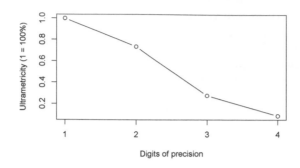

FIGURE 5.2: Dependence of ultrametricity (i.e. data inherently hierarchical) on precision; 20,000 chemicals used. See Section 5.8.2 for details.

measure, which we term α, is the number of ultrametricity-verifying triangles divided by the total number of triangles. When all triangles respect the ultrametric properties this yields $\alpha = 1$; if no triangle does, then $\alpha = 0$. For N objects, exhaustive enumeration of triangles is computationally prohibitive, so we sample i, j, k in practice. We sample uniformly over the object set.

At least two other approaches to quantifying ultrametricity have been used, but in [169] we point to their limitations. Firstly, there is the relationship between subdominant ultrametric and given dissimilarities (see [207]). Secondly, we may look at whether the interval between median and maximum rank dissimilarity of every set of triplets is nearly empty. Taking ranks provides scale invariance. This is the approach of [137].

A further interesting development, based on graph properties that encompass topological relationships, is described in Annex 2 of Chapter 9 below.

5.8.2   Pervasiveness of Ultrametricity

As noted above, in Section 5.1, we find experimentally that ultrametricity is pervasive in the following ways (see [92, 169]). As dimensionality increases, so does ultrametricity. In very high-dimensional spaces, the ultrametricity approaches 100%. Relative density is important: high dimensionality and spatial sparsity mean the same in this context.

Compared to metric embedding, and model fitting in general, mapping metric (or other) data into an ultrametric, or embedding a metric in an ultrametric, leads to study of distortion.

As such a distortion, let us look at recoding of data by modifying the data precision. By a focus on the data measurement process, we can find a new way to discover (hierarchical) structure in data.

In Figure 5.2, 20,000 encoded chemicals were used, normalized as described in [186]. Next, 2000 sampled triangles were selected, and ultrametricities obtained for precisions $1, 2, 3, 4, \ldots$ in all values. Numbers of non-degenerate triangles (out of 2000) were found as follows (where non-degenerate means isosceles with small base):

- precision 1: 2

- precision 2: 1062

- precision 3: 1999

- precision 4: 2000.

Thus if we restrict ourselves to just 1 digit of precision we find a very high degree of ultrametricity, based – to be noted – on the preponderance of equilateral triangles. With 2 digits of precision, there are a lot more cases of isosceles triangles with small base.

Thus we can bring about higher ultrametricity in a data set through reducing the precision of data measurement.

5.9 Generalized Ultrametric and Formal Concept Analysis

5.9.1 Generalized Ultrametric

The usual ultrametric is an ultrametric distance, that is, for a set I, $d : I \times I \longrightarrow \mathbb{R}^+$. The range is the set of non-negative reals.

The generalized ultrametric is $d : I \times I \longrightarrow \Gamma$, where Γ is a partially ordered set, or poset. In other words, the *generalized* ultrametric is a set. The range of this generalized ultrametric is therefore defined on the power set or join semilattice. The minimal element of the poset generalizes the zero distance of the mapping onto $\mathbb{R}^+$ case.

Comprehensive background on ordered sets and lattices can be found in [56]. A review of generalized distances and ultrametrics can be found in [215]. Generalized ultrametrics are used in logic programming and, as we will discuss in the subsection to follow, formal concept analysis can be seen as use of the generalized ultrametric.

To illustrate how the generalized ultrametric is a vantage point on the Baire framework, we focus on the common, or shared, prefix aspect of this. Consider the following data values: $x_1 = 0.4573$, $x_2 = 0.4843$, $x_3 = 0.45635$, $x_4 = 0.4844$, $x_5 = 0.4504$. Common prefixes are as follows, where we take decimal digits (i.e. "4573" in the case of x_1):

$d(x_1, x_2) = 4$
$d(x_1, x_3) = 4, 45$
$d(x_1, x_4) = 4$
$d(x_1, x_5) = 4, 45$
$d(x_2, x_3) = 4$
$d(x_2, x_4) = 4, 48, 484$
$d(x_2, x_5) = 4$
$d(x_3, x_4) = 4$
$d(x_3, x_5) = 4, 45$
$d(x_4, x_5) = 4.$

The partially ordered set is just the structure with the "subset of the common prefix" binary relation with, on one level, the single-valued common prefixes (here 4); the next level has the common prefixes of length 2 (here 45, 48); the following level has the common prefixes of length 3 (here 484). Prior to the first level, we have the common prefix of length 0 corresponding to no match between the strings.

Thus we see how we can read off a partially ordered set, contained in a lattice, based on the common or shared prefixes.

5.9.2 Formal Concept Analysis

In formal concept analysis (FCA) [56], we focus on the lattice as described in the previous subsection. Thus, we can say that a lattice representation is yet another way of displaying (and structuring) the clusters and the cluster assignments in our Baire framework. For more discussion of FCA, see Section 3.4.

Following a formulation by Mel Janowitz (see [101, 102]), lattice-oriented FCA is contrasted with poset-oriented hierarchical clustering in the following way, where by "summarize" is meant that the data structuring through lattice or poset allows statements to be made about the cluster members or other cluster properties.

- Cluster, then summarize. This is the approach taken by (traditional) hierarchical clustering.

- Summarize, then cluster. This is, in brief, the approach taken by FCA.

Further description of FCA is provided in [185]. Our aim here has been to show how the common prefixes of strings lead to the Baire distance, and also to a generalized ultrametric, and furthermore to a poset and an embedding in a lattice.

5.10   Linear Time and Direct Reading Hierarchical Clustering

5.10.1   Linear Time, or $O(N)$ Computational Complexity, Hierarchical Clustering

A point of departure for our work has been the computational 'objective of bypassing computationally demanding hierarchical clustering methods (typically quadratic time, or $O(N^2)$ for N input observation vectors), but also having a framework that is of great practical importance in terms of the application domains.

Agglomerative hierarchical clustering algorithms are based on pairwise distances (or dissimilarities) implying computational time that is $O(N^2)$ where N is the number of observations. The implementation required to achieve this is, for most agglomerative criteria, the nearest neighbour chain, together with the reciprocal nearest neighbours, algorithm (furnishing inversion-free hierarchies whenever Bruynooghe's reducibility property (see [167]) is satisfied by the cluster criterion).

This quadratic time requirement is a worst-case performance result. It is most often the average time also since the pairwise agglomerative algorithm is applied directly to the data without any preprocessing speed-ups (such as preprocessing that facilitates fast nearest neighbour finding). An example of a linear average-time algorithm for (worst-case quadratic computational time) agglomerative hierarchical clustering is given in [162].

With the Baire-based hierarchical clustering algorithm, we have an algorithm for linear-time, worst-case hierarchical clustering. It can be characterized as a divisive rather than an agglomerative algorithm.

5.10.2   Grid-Based Clustering Algorithms

The Baire-based hierarchical clustering algorithm has characteristics that are related to grid-based clustering algorithms and density-based clustering algorithms which often were developed in order to handle very large data sets.

The main idea here is to use a grid-like structure to split the information space, separating the dense grid regions from the less dense ones to form groups. In general, a typical approach within this category will consist of the following steps [91]:

1. Creating a grid structure, i.e. partitioning the data space into a finite number of non-overlapping cells.

2. Calculating the cell density for each cell.

3. Sorting of the cells according to their densities.

4. Identifying cluster centres.

5. Traversal of neighbour cells.

Additional information about grid-based clustering can be found in [47, 85, 202, 248].

In Sections 5.6 and 5.7 in particular it has been described how cluster bins, derived from an m-adic tree, provide us with a grid-based framework for data structuring. We can read off the cluster bin members from such an m-adic tree. In Section 5.10.1 we have noted how an m-adic tree requires one scan through the data, and therefore this data structure is constructed in linear computational time.

In such a preprocessing context, clustering with the Baire distance can be seen as a "crude" method for getting clusters. After this we can use more traditional techniques to refine the clusters in terms of their membership. Alternatively (and we have quite extensively compared Baire clustering with, for example, k-means, where it compares very well), clustering with the Baire distance can be seen as fully on a par with any optimization algorithm for clustering. As optimization, and just as one example from the many examples reviewed in this chapter, the Baire approach optimizes an m-adic fit of the data simply by reading the m-adic structure directly from the data.

5.11 Summary: Many Viewpoints, Various Implementations

Baire distance is an ultrametric, so we can think of reading off observations as a tree.

Through data precision of measurement alone we can enhance inherent ultrametricity, or inherent hierarchical properties in the data.

Clusters in such a Baire-based hierarchy are simple "bins" and assignments are determined through a very simple hashing (e.g. $0.3475 \longrightarrow$ bin 3, and $\longrightarrow$ bin 34, and $\longrightarrow$ bin 347, and $\longrightarrow$ bin 3475).

Observations are mapped onto sets (e.g. 0.3475 and 0.3462 are mapped onto sets labelled by 3 and 34). We therefore have a generalized ultrametric. A lattice can be used to represent range sets, leading also to a link with formal concept analysis.

Our wide range of vantage points on the Baire-based processing is important because of the many diverse applications in view, including the structuring of the data, the reading off of clusters, the matching or best fit of observations, and the determining of hierarchical properties.

Apart from showing how the Baire vantage point gives rise in practice to such a breakthrough result of having linear-time hierarchical clustering, the other important contribution in this approach has been to show how so many vantage points can be adopted in this way on data, on the structuring and embedding of data, and ultimately on the interpretation and exploitation of data.

6

Big Data Scaling through Metric Mapping

In this chapter, the remarkable simplicity of very high-dimensional spaces is exploited. There is a particular focus on correspondence analysis as a geometric approach. In the examples used, there are dimensionalities up to around 1 million. A particular benefit of correspondence analysis is its suitability for carrying out an orthonormal mapping, or scaling, of power law distributed data. Power law distributed data are found in many domains. Correspondence factor analysis provides a latent semantic or principal axes mapping. The examples use data from digital chemistry and finance, and other statistically generated data. Applications are then discussed in regard to analysis in high dimensions of social media, and general domains where the objective is efficiently to induce the hierarchical clustering of the data.

Points in high-dimensional spaces become increasingly equidistant as dimensionality increases, as demonstrated experimentally in [169]. Gaussian clouds in very high dimensions are studied in [92, 65]. The latter finds that "not only are the points [of a Gaussian cloud in very high-dimensional space] on the convex hull, but all reasonable-sized subsets span faces of the convex hull. This is wildly different than the behavior that would be expected by traditional low-dimensional thinking."

That very simple structures come about in very high dimensions can have far-reaching implications. Firstly, even very simple structures (hence with many symmetries) can be used to support fast and perhaps even constant-time worst-case proximity search [169]. Secondly, as shown in the machine learning framework by [92], there are important implications ensuing from the simple high-dimensional structures. Thirdly, [175] shows that very high-dimensional clustered data contain symmetries that in fact can be exploited to "read off" the clusters in a computationally efficient way. Fourthly, following [58], what we might want to look for in contexts of considerable symmetry are the "impurities" or small irregularities that detract from the overall dominant picture.

In general, data analysis, considered as the search for symmetries in data, is discussed in Chapter 3. This relates in particular to hierarchical clustering.

Traditionally, operations are considered in a space endowed with a given metric. For example, bounds are established on a projection operator in a Hilbert space, leading to what is referred to as a frame synthesis operator, which satisfies the generalized Parseval relation. This is discussed in [224, p. 45]. Such methodology is often concerned with sparsifying data. That is, the transformed data are far sparser than the original data. That helps greatly with data compression, or object and feature finding. Next, we will make note of the methodology for approximating distances in a lower-dimensional space embedding. The embedding is to be a minimum-distortion one. Bounds for projected distances relative to given distances are provided for in the Johnson–Lindenstrauss lemma [55].

Here, however, our overriding objective is not so much fidelity in a new data space relative to our original data. Rather, we are concerned with data scaling. This takes the following form. Our given data clouds are endowed with an appropriate metric for positive counts or other measurements, the χ^2 metric. We map the data clouds into space that is endowed with the Euclidean metric. Distance invariance holds between χ^2 distance in input space and Euclidean distance in the output, factor space.

The mapping criterion respects scaling properties. We may express this as follows: an observable (row of our data matrix) is mapped into a factor (or latent semantic) space, such that it is positioned at the centre of gravity of all its attributes (column vectors of our data matrix).

The general context, some of the properties of very high-dimensional spaces, is examined and the implications that follow (Section 6.7). Background on the scaling or mapping methodology is provided (Section 6.8). Findings of this are reported on (Sections 6.9–6.11). Consequences for applications are described (Sections 6.12 and 6.13). An Annex to this chapter provides software code which will allow easy reproducibility of results. See the book's website for easy access to this code. Regarding notation mainly used (cf. [28]), it is a simple form of Einstein tensor notation. At all times, the terms used are clearly defined.

6.1   Mean Random Projection, Marginal Sum, Seriation

The mean random projection approximates the marginal sum and will be used for seriation, or a one-dimensional mapping, that will then be used as a basis for the clustering. There is little difference from random projections that are normalized.

Our given data are here denoted x_{IJ}, that is, $\{x_{ij}|i \in I, \ j \in J\}$, $n = |I|$, $m = |J|$. As a random projection, we use values uniformly distributed in $[0, 1]$, denoted u_J^1, that is, $\{u_j^1|j \in J\}$. The superscript is 1 because this is our first random axis. A random projection is the matrix–vector product $r_I^1 = x_{IJ} \ u_J^1$. Another random axis is generated, and the projection on it is determined: $r_I^2 = x_{IJ} \ u_J^2$. A third random axis is generated, and the projection on it is determined. This continues in this particular descriptive context for 100 random projections. Thus we have random projections $r_I^1, r_I^2, r_I^3, \ldots, r_I^{100}$.

Our interest is in the mean of the random projection vectors. The motivation for this is that if the random projections are well correlated, as we find to be the case for high-dimensional data clouds, then the mean is a suitable consensus. So we look at the succession of mean projection vectors: for $K = 1, 2, \ldots, 100$, $\frac{1}{K} \sum_{k=1}^{K} r_I^k$. The random axis used on each occasion, $k = 1, 2, \ldots, K$, is newly generated.

The marginal sums, $x_I = \{\sum_{j=1}^{m} x_{ij}|i \in I\}$ are a constant times the mean vector. We can expect the mean random projection to approximate the cloud mean very well. The cloud is the point set x_{iJ}, for $i \in I$. That is to say, we have (1) the cloud's mean vector, and (2) the random cloud that consists of randomly generated points that are a linear combination of the cloud's given points.

The uniformly distributed random axes result in the projection values being nonnegative, when the initial given data are non-negatively valued. To set the context for this approach, let us consider as a case study with a very small data set a 25×12 data array of values uniformly distributed in $[0, 1]$. (We just note that this random generation was different from other random generations, such as for the random axes.) Then a second study uses another initial 25×12 data array of Gaussian distributed values, of mean 0 and standard deviation 1. Since some of these values are negative, the random projections, using the uniformly distributed axes, consist of negative as well as positive projection values.

Due to the following, we can note that input data values are not constrained to be, for example, non-negative in value. We note this because in correspondence analysis, we are dealing with marginal distributions that are mass distributions of row (observations, I) cloud and column (attributes or properties, J) cloud. Thus in correspondence analysis, we must consider non-negative-valued masses, generally also non-negative given data. In the current case study, we have the following: the random projection clouds are resulting

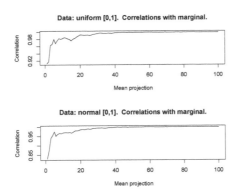

FIGURE 6.1: Pearson correlations between mean random projection, and the marginal sums. The two input data tables used were of dimensions 25×12. Maximum and minimum values of the input data tables are, respectively, $0.9991, 0.0012$ and $2.6479, -3.0277$.

from a linear combination of of the point cloud's given points; and the marginal sum is a constant times the mean marginal. Therefore it is not relevant whether we are working with non-negative-valued data, data consisting of both positive and negative values, or of course, positive-valued data.

Figure 6.1 demonstrates clearly how the mean random projection (with uniformly distributed random axes) very well approximates the marginal sum, for a sufficient number of random axes. The marginal sum is proportional to the mean of the point cloud, $I \subset \mathbb{R}^m$. Here, the point cloud's coordinates are real-valued, in the space of dimensionality $m = |J|$. To illustrate the degree of approximation that is displayed in Figure 6.1, the final three correlation values for the two data sets used are respectively $0.9985, 0.9986, 0.9984$ and $0.9986, 0.9986, 0.9985$. That is to say, these are correlations between the marginal row sums and the means of 98, 99 and 100 random projections.

This study will serve as important background for our further work. To summarize the outcome: *the mean of 40 or more projections of the point cloud on uniformly distributed axes approximates very well the marginal sum of the point set.* The property of 40 or more projections is observed in Figure 6.1 and in many other experiments. The projections on uniformly distributed axes are of the point cloud, which is not centred, nor normalized in any way. The axes being uniformly distributed means that these axes are defined as uniformly distributed coordinates in the given real m-dimensional space, $\mathbb{R}^m$. The marginal sum of the point set is directly proportional to the means of the point coordinates. That is, each point's marginal sum is $x_i = \sum_{j=1}^{m} x_{ij} = m \frac{1}{m} \sum_{j=1}^{m} x_{ij} = m$ times the mean of the coordinate values of x_i.

6.1.1 Mean of Random Projections as A Seriation

The mean of the random projections generating a seriation is to be used for subsequent clustering. In this subsection, the robustness and general applicability of this is demonstrated through the limited differences between parametric and non-parametric correlations.

In Figure 6.2, a non-parametric correlation is used. This is in order to assess rank-order properties that may be particularly appropriate for clustering data, in a consistent way, using the seriation obtained. The non-parametric correlation, here Kendall's coefficient of correlation, replaces the Pearson correlation used Figure 6.1. In other respects, this Kendall

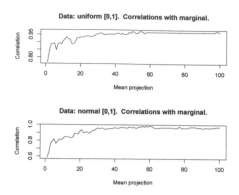

FIGURE 6.2: Kendall rank correlations between mean random projection and the marginal sums. The two input data tables used were of dimensions 25×12. Maximum and minimum values of the input data tables are respectively $0.9991, 0.0012$ and $2.6479, -3.0277$.

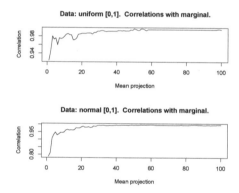

FIGURE 6.3: Spearman rank correlations between mean random projection and the marginal sums. The two input data tables used were of dimensions 25×12. Maximum and minimum values of the input data tables are, respectively, $0.9991, 0.0012$ and $2.6479, -3.0277$.

rank-order correlation study uses the same data and the same processing steps as in the previous section (with the same seeding of the random number generation).

We find in Figure 6.2 that the approximation of the mean projection, by the marginal sums, is not quite as accurate (compared to the results obtained using the Pearson correlation, as displayed in Figure 6.1). For our two input data sets we find correlations for the final three means of projections, that is, for means of $k = 98, 99, 100$ random projections, of $0.973, 0.973, 0.967$ (corresponding to the top graph in Figure 6.2), compared to $0.973, 0.973, 0.973$ (corresponding to the bottom graph in Figure 6.2).

Figure 6.3 shows the same processing outcome using Spearman's rank-order correlation. Correlations for the final three means of projections are $0.996, 0.996, 0.995$ (corresponding to the top graph in Figure 6.3), compared to $0.997, 0.997, 0.997$ (corresponding to the bottom graph in Figure 6.3). In the latter case, it can be noted that the fourth last value of the mean projections (i.e. for 97 random axes), is not the same in value relative to 98, 99, and 100 random axes, but is found to be 0.996.

We find the Pearson correlation to be the highest in value, followed by the Spearman

	$k = 98$	$k = 99$	$k = 100$
Pearson	0.9985497	0.9985972	0.9984356
Spearman	0.9961538	0.9961538	0.9953846
Kendall	0.9733333	0.9733333	0.9666667
Pearson/max. normalized	0.9985769	0.9986095	0.9984502
Pearson/norm normalized	0.9985652	0.9985888	0.9984162

TABLE 6.1: Correlation between mean random projection and row sum for 98, 99 and 100 random projections. Parametric and non-parametric correlations are used. In the normalized cases, both the mean of the projections and the row sum are normalized to have maximum value 1 and L_2 norm 1.

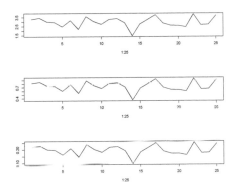

FIGURE 6.4: Mean random projection (top to bottom) corresponding to rows 1, 4, 5 of Table 6.1 (i.e. non-normalized, normalized by maximum value (L_∞ or Chebshev metric), projections normalized to unit norm). Input data uniformly distributed, of dimensions 25 × 12.

correlation, followed in turn by the Kendall correlation. Nonetheless, the approximation is very good. For Pearson and Spearman, it is observed to be, with sufficient random axes, equal to or greater than 0.99 in all cases. For the Kendall correlation, it is around 0.97.

6.1.2 Normalization of the Random Projections

Much similarity is found between outputs that have had differing normalizations of the random projections. Further assessment is carried out for normalizations using the L_∞ (maximum or Chebyshev) norm and the L_2 norm. Table 6.1 shows very little difference between the approaches used.

The correlations in all cases are similar, as the means of the random projections are considered, relative to the row sums. So therefore let us look at the final mean random projection, for $k = 100$. That is therefore the mean of 100 random axes. Each random axis is uniformly distributed. For the normalized random projection values, the scale will differ depending on the normalization used. Nonetheless, we do observe a very similar outcome (see Figure 6.4).

While this has been a small study, it is used for illustrative purposes.

6.2   Ultrametric and Ordering of Rows, Columns

In this section, let us use previously displayed outputs from Chapter 3. Consider the clustering of the data to start the description of clustering properties of the marginal sums, which, as has been shown, is well approximated by the mean random projection.

The hierarchy in Figure 3.2 has the ultrametric or tree-originating distances in Table 3.2. From that table the graphical presentation or visualization displayed in Figure 3.2 follows. The latter figure has the following ordering of row and column labels, relating to the seven iris flowers here. This ordering or permutation, which is identical for the rows and the columns given the symmetry of the ultramatric matrix, is related to the order or permutation of the terminal nodes in the hierarchy in Figure 3.2. Figure 3.2 (left) has the terminal nodes in the order $2, 7, 3, 4, 6, 1, 5$. If the left child subnode precedes agglomeration with the fight child node, then the order of terminal nodes is $3, 4, 7, 2, 1, 5, 6$. In [176], there is discussion of row and column permuting in the original data table, to lead directly to such output display. That is relevant here because the row (or column) sums of the ultramatric matrix (i.e. cophenetic distance matrix) express this order in this case.

In [137, 176], and as described in Chapter 3, the following is the specification of the block diagonal structure that respects the ultrametric inequality. For a symmetric positive definite matrix, with values $d(i, j)$, for all i, j, k, the ultrametric inequality is $d(i, j) \leq \max(d(i, k), d(j, k))$. For an $n \times n$ matrix of positive reals, symmetric with respect to the principal diagonal, to be a matrix of distances associated with an ultrametric distance on X, a sufficient and necessary condition is that a permutation of rows and columns satisfies the following form of the matrix:

1. Above the diagonal term, equal to 0, the elements of the same row are non-decreasing.

2. For every index k, if

$$d(k, k+1) = d(k, k+2) = \cdots = d(k, k+\ell+1)$$

then

$$d(k+1, j) \leq d(k, j), \quad \text{for } k+1 < j \leq k+\ell+1,$$

and

$$d(k+1, j) = d(k, j), \quad \text{for } j > k+\ell+1.$$

Under these circumstances, $\ell \geq 0$ is the length of the section beginning, beyond the principal diagonal, the interval of columns of equal terms in row k.

In approximating a potentially valid seriation of what is to be clustered, we are seeking a permutation that is consistent with what has been described in this section. A more relevant permutation that would be directly associated with the dendrogram would be $3, 4, 7, 2, 1, 5, 6$ (see Figure 3.2). Note how this is identical to the order of row (identically column) sums of the ultrametric matrix.

6.3   Power Iteration Clustering

Whether correspondence analysis or any other eigendecomposition, the processing can be carried out iteratively as follows. For a matrix, A, we seek to solve $Au = \lambda u$. The solution is the first (largest) eigenvalue, λ, associated with the eigenvector, u. Choose a random, non-zero, initial vector, t_0. Then define $t_1, t_2, \ldots$ as follows:

$$At_0 = x_0, \quad t_1 = x_0/\sqrt{x_0' x_0},$$
$$At_1 = x_1, \quad t_2 = x_1/\sqrt{x_1' x_1},$$
$$At_2 = x_2, \quad t_3 = x_2/\sqrt{x_2' x_2},$$
$$\ldots$$

The normalization to unit L_2 norm can be viewed as preventing the vector t_k from getting too large [143]. However, in [143] use the L_1 norm is used. A further justification for the L_2 norm is principal components analysis or correspondence analysis where the factor space, defined by the eigenvalue and eigenvectors, is endowed with the Euclidean, L_2 metric.

We halt the iterations where there is convergence: $|t_n - t_{n+1}| \leq \epsilon$. At convergence, we have the approximation of $t_n = t_{n+1}$. Therefore, $At_n = x_n$. Since $t_{n+1} = x_n/\sqrt{x_n' x_n}$ we may substitute terms in $At_n = x_n$ to give $At_n = \sqrt{x_n' x_n} t_{n+1}$. Since $t_n = t_{n+1}$, that allows us to conclude that t_n is the eigenvector, and the associated eigenvalue is $\sqrt{x_n' x_n}$.

We can partial out the first eigenvector and eigenvalue in order to proceed to the next eigenvector and eigenvalue (we redo the analysis on the matrix, $A_{(2)} = A - \lambda u_1 u_1'$). We will not do this here since our interest is in the first eigenvector. This iterative solution is described in [191, Section 2.2.6].

In [143], this approach is used with a data table that is normalized through division by the row sum. This is the row (object) profiles. For a data table $x_{IJ} = \{x_{ij} \mid i \in I, \ j \in J\}$, the profile is $\{x_{ij}/x_i \mid i \in I, \ j \in J\}$, where $x_i = \sum_{j \in J} x_{ij}$. In practice, frequencies are used: $f_{ij} = x_{ij}/\sum_{i \in I, j \in J} x_{ij}$. Similarly, we have $f_i = x_i/\sum_{i \in I} f_i$. In tensor notation (Einstein tensor notation, used in [24]), the row profile is $f_J^i = \{f_{ij}/f_i \mid j \in J\}$. Now, working with row profiles means that the row sums are constant, $\sum_{j \in J} f_{ij}/f_i = 1$. In seeking to solve $f_J^I u_J = \lambda u_J$ we have that $u_J = 1_J$ where the latter term is a $|J|$-length vector of 1s. The expression we want to solve then requires $\lambda = 1$. This is the so-called trivial first eigenvector and eigenvalue in correspondence analysis. It is due to the centring of the cloud of row points and column points. Since this is trivial in value, [143] present the case for iterating towards the solution but stopping at a local optimum in regard to convergence. Rather than the first eigenvector, an "intermediate vector" is obtained. The iterative scheme "converges locally within a cluster".

Yan et al. [249] note how power iteration clustering can be based on the dominant eigenvalue/eigenvector described above. That points to the relevance of the data array used. It has been noted above how row profiles were used in [143]. In [249], reference is made in particular to spectral clustering, described as "a family of methods based on the eigendecomposition of affinity, dissimilarity or kernel matrices". Such spectral clustering is core to correspondence analysis which, typically in practice, permits the inducing of a hierarchical clustering, for example, from the factor space endowed with the Euclidean metric. In [143], k-means clustering (partitioning) is used.

Further application of [143] is carried out in [145], where by means of the dominant eigenvector found using the power iteration eigenreduction approach, both row and column sets are used. This allows row and column (of the data array) permutation to yield a block clustered view of the data array. A wide-ranging review of row and column permutation for clustering is given in [142].

In [145], comprehensive use is presented of row and column permuting, using power iteration clustering [143] applied successively to the rows and to the columns. The input data array used in [145] is the sums of squares and cross-products matrix; that is, for the initial (non-negative dependency values, described in graph terms) matrix, X, the input for row and column permutation is $X'X$.

6.4   Input Data for Eigenreduction

Jean-Paul Benzécri [24] describes his synoptic viewpoint, across the board on data analytics, as follows: "All in all, doing a data analysis, in good mathematics, is simply searching eigenvectors; all the science (or the art) is just to find the right matrix to diagonalize."

For correspondence analysis, we have the following. In [29], Benzécri discusses analysis of data sets that are unbounded in number of rows (an infinite set), with possibly 1000 columns. For given data k, and factor G on the set J, the eigenvalue and eigenvector decomposition is the solution of the equation

$$\sum_{j' \in J} \left(\left(\sum_{i \in I} \frac{x_{ij'}}{x_i} \frac{x_{ij}}{x_i} \right) G_\alpha(j') \right) = \lambda_\alpha G_\alpha(j). \tag{6.1}$$

That, therefore, relates to the column profiles. Following determination of the factors G on J, the transition formula relationship allows the determination of the factors F on I.

The full eigenreduction, determining the full set of factors, is described in [29, 30]. In the latter, there is discussion of the extensive use by Ludovic Lebart of stochastic approximation of factors and associated contributions to inertia, defined from the eigenvectors and associated eigenvalues. Lebart et al. [134, Chapter VI] provide a comprehensive description, with this eigenreduction termed a "direct reading" implementation. In [29], since the set of eigenvectors and associated eigenvalues is the objective, a matrix of trial vectors is initialized and converged, in the sequence of iterations, to the desired outcome (see [29, equation (6), p. 391]). This is as in the approach employed in [50].

In [28], there is a short review of an alternative approach by finding subtables of the given input data table (see Section 6.2 above). For a subset J_s of J, analyses are carried out on these $I \times J_s$ subtables. This work was primarily due to Brigitte Escofier. For one $I \times J_s$ subtable, there can be added (juxtaposed) a "rest" column with the accumulation of all columns in the set $J - J_s$. In this case, one has a Euclidean representation (i.e. the factor space) of the cloud $N(J)$ that is the same as the analysis carried out on the $I \times J$ table, but relative to axes that are adjusted to the subcloud, $N(J_s)$. While the full analysis can be reassembled from the analyses of the subtables, it may be the case that the subtable analyses are more interesting in their own way. The global analysis, it is stated, may be perturbed by such singular modalities of variables, in particular missing values. The views of $N(J)$, adjusted to well-chosen subclouds $N(J_s)$, can better show the global structure.

For principal components analysis, depending on the normalization applied, the matrix to be diagonalized is the sums of squares and cross-products matrix, the variance–covariance matrix, or the correlation matrix. See, for example, [191] for discussion of these inputs for PCA.

6.4.1   Implementation: Equivalence of Iterative Approximation and Batch Calculation

It has been noted in Section 6.3 how a limited number of iterations may be used, rather than convergence. Let us also look now at where and how a fixed number of iterations may be beneficial, as an alternative to convergence.

We use the following illustrative and motivational example from [29]. This example relates to estimating the mean of the (potentially unbounded) set of values $x_1, x_2, x_3, \ldots, x_n, \ldots$. If the number of such values is known (and denoted N), then that leads to just determining the estimated mean on all the data. If there were weights involved, then an unbounded sequence becomes more problematic. Now, it can be shown that two successive values in the estimation have the relationship $\mu_{n+1} = \mu_n + ((x_{n+1} - \mu_n)/(n+1))$. It is seen that the $(n+1)$th value can be considered as correcting the estimate at that iteration. Also we see that if $\mu_{n+1} = \mu_n$, then that correction to the estimate at that iteration would be equal to 0. (It is acknowledged in [29] that successive updates, carried out in this way, may lead to accumulation of rounding errors. On the other hand, it is considered that any exceptional or outlying value of x would be very clearly indicated.)

In practice, the iterative estimation can be useful and relevant as an approach to determining the mean, especially of an unbounded sequence. An assumption, to be considered in each case, is the underlying distribution of the x terms.

We conclude the following from this small, illustrative case study. Given an underlying distribution, we can either (i) iterate until convergence, or (ii) assume a fixed value of N and carry out the computation for the sequence of values (or vectors) that are from 1 to N.

6.5 Inducing a Hierarchical Clustering from Seriation

Consider now the inducing (i.e. generation) of a hierarchical clustering from seriation (i.e. from a unidimensional representation of our observations). The following is based on [53], which establishes the foundations for inducing a hierarchical clustering from a newly represented, or newly encoded, mapping of our data. This very important result allows us to seek a seriation in order to hierarchically cluster our data in a computationally very efficient manner.

Consider a dendrogram and the terminal nodes in a sequence, $\pi(I)$, a permutation of the object set, I. For $i \in I$, $i = 1, 2, 3, \ldots$, $n = |I|$, consider π_i (i.e. $\pi_1, \pi_2, \ldots, \pi_n$). Now define the $(n-1)$-vector, t, with general element $t_j = x_{j+1} - x_j$. Such a one-dimensional ordering, t, is compatible with the given dendrogram ordering if $j \leq k$ implies $x_j \leq x_k$. Then we define a matrix of interpoint distances in the unidimensional ordering as follows: $d_{jj} = 0; j < k \longrightarrow d_{kj} = d_{jk} = \sum_{l=j}^{k-1} t_l$.

As noted in [30], although quite likely to be fully justified in practice, any iterative refinement algorithm is unable to deliver an optimal solution. The non-uniqueness of the seriation or unidimensional scaling that can be the starting point for inducing a hierarchical clustering is a limitation in practice, since many alternatives may (or may not) be relevant for the hierarchy to be induced.

Using our approach on the Fisher iris data [76], 150 flowers crossed by petal and sepal width and breadth, provides the following outcome. We determine row sums of the initial 150×4 data matrix, and the mean random projection of projects on 100 uniformly generated axes. From our previous results, we know that these are very highly correlated. We construct hierarchical clusterings on (i) the original 150×4 data matrix, (ii) the mean random projection, and (iii) the row sums. The cophenetic correlation coefficient is determined (this uses ultrametric distances derived from the hierarchical tree, or dendrogram). We find the cophenetic correlation of the hierarchies constructed on the row sums, and on the mean random projection, to be equal to 1 (as anticipated). Then between the hierarchy constructed on the 150×4 data matrix, and the mean random projection, the cophenetic correlation coefficient is 0.8798. For the given data and the row sums, it is 0.9885. The hierarchical clustering used was the average method; and other methods, including single link, provided very similar results. The distance used as input to the hierarchical agglomerative clustering was the square root of the squared Euclidean distance. Other alternatives can be, and have been, looked at, from the point of view of the distance used, and from the point of view of the agglomerative hierarchical clustering criterion.

All in all, these results set the overall perspectives to the applications to be described below, in later sections of this chapter.

We also looked at uniformly distributed (on $[0, 1]$) data of dimensions 2500×12. The correlation between row sums and the mean of 100 random projections was 0.99. However, the correlation between the hierarchical clustering on the original data, and the mean random projection, was 0.58. The correlation with the row sums was 0.578. The performance on these randomly generated data is seen to be not as good as that on the real-valued Fisher data. But it must be noted that such uniformly distributed data, by construction, are not intended to contain non-trivial clusters.

We conclude this practical case study with the following remarks. Firstly, for real data, we found a very good result. Secondly, the lack of uniqueness of the seriation (or unidimensional representation) means that various possibilities may, or may not, be most appropriate. But, finally, we determined a very good outcome, which very much respected (correlation between induced hierarchical clusterings of around 0.88) the clustering properties in the data.

6.6   Short Summary of All These Methodological Underpinnings

We have set out the following objectives, algorithms and implementations. In all cases important properties and characteristics have been discussed.

1. The use of iterative approximation, to convergence; to optimally summarize the data cloud by means of the dominant eigenvector.

2. The data cloud centroid, for its role in summarizing the cloud, and relationships with the dominant eigenvector.

3. We have referred to importance for applications of instances of data cloud piling and concentration.

4. How such summarization induces seriation of our data cloud.

5. General clustering properties of such a seriation. This is motivated by the awareness that an ultrametric topological embedding of our data cloud in a space of arbitrary dimensionality (alternatively expressed: a hierarchical structuring of our data) can be perfectly scaled in one dimension.

6. Noting the importance of the following: no approximate scheme can be guaranteed to provide an optimal outcome; selection of data table normalization plays an important role; convergence, consistency and stability properties of implementations.

As noted in [30], any approximate scheme cannot be guaranteed to provide the optimal outcome. This is simply noting what approximation means.

6.6.1   Trivial First Eigenvalue, Eigenvector in Correspondence Analysis

Following a short recap of processing, the role of the first eigenvalue, trivial since it is equal to vectors of 1s, is noted. This eigenvalue and its associated trivial eigenvector, are removed

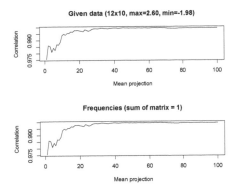

FIGURE 6.5: Similar to Figure 6.1, but with negative values in the given data.

in practice from results obtained from the essential, eigenreduction, processing. This section is a set of notes in regard to this.

We consider our given data, x_{IJ}, as non-negative values. Convert to frequencies, f_{IJ}, such that the overall sum is 1. The row marginals of given data and frequencies are, respectively, x_I and f_I . The mean of the random projections converges to the row marginal. We plot the correlations between $1, 2, 3, \ldots, 100$ random projection means versus the row marginal. We find no difference between original data and frequencies (this is very clear, namely that division by a constant will not affect the relative relationships). We even find that this holds for original data containing some negative values. Eigenreduction is of the matrix of (j, k) term: $\sum_i x_{ij} x_{ik} / x_i x_k$.

The trivial (first, and never used in practice) eigenvalue is 1, and the associated eigenvector has all 1 or -1 values. Row contributions on this trivial factor are mass times squared distance from the origin. For the first trivial (unused in practice) factor, we therefore have that the contributions to its inertia by the row set are identical to the marginal values, or row sums.

Figure 6.1 shows convergence towards perfect positive correlation of the mean of the random projections *vis-à-vis* the row marginal sums, which are the row contributions to the inertia of this first (trivial) axis, in the correspondence analysis – factor analysis – framework. Figure 6.5 is similar, but here the input data have negative values, and are seen to be fully effective.

6.7 Very High-Dimensional Data Spaces: Data Piling

With high-dimensional, sparse data [92], there is a very strong concentration of our clouds (rows/points, columns/projection vectors) into concentrated (i.e. small variance) Gaussians. Therefore, there is a good approximation of our cloud by its mean. This in turn means that the mean random projection is a very good representative of our data cloud.

Our data are points in a high-dimensional data cloud. In correspondence analysis, such points are *profiles*, as will be described below in Section 6.8.1.

Through high-dimensional piling (i.e. concentration), we have that the profile vectors tend towards the average profile. What gives rise to this is sparsity through high dimensionality, which also implies low sample (or population) size. It implies this because we are not considering here the case of both population size and dimensionality tending to infinity at the same, or related, rate.

By the central limit theorem, and by the concentration (data piling) effect of high dimensions [92, 230], we have, as dimension $m \to \infty$, that pairwise distances become equidistant, and orientation tends to be uniformly distributed. We find also that the norms of the target space axes are Gaussian distributed; and, as typifies sparsified data, the norms of the points in our high-dimensional data cloud, in the factor space, are distributed as a negative exponential or a power law.

6.8 Recap on Correspondence Analysis for Following Applications

This section is largely another presentation of what is covered in the annex to Chapter 1.

In general, correspondence analysis is used with qualitative (or categorical) data, or mixed qualitative and quantitative data. It is fully applicable to non-negative quantitative data. In brief, principal components analysis typically works on (i) sums of squares and cross-product values, implying that no normalization is carried out to the input data; (ii) variances and covariances, that result from normalizing the input data by centring the variables or attributes; and (iii) correlations, that result from standardizing the input data, through centring the variables and also reducing them to unit variance. Correspondence analysis normalizes the input data by taking account of the marginal distributions of both rows and columns. This will be further described below. As will be seen, this is a normalization resulting from division by a sum of (row or column) values. Alternative analysis options in correspondence analysis are available at the early stage of the selection of data to analyse. This can include categorical coding of quantitative data, or selection of categorical modalities. See [171] for an in-depth discussion of correspondence analysis in terms of what measurements are used, and how they are encoded, prior to analysis.

For input data that cross rows i by columns j, and where f_{ij} is the corresponding frequency value, the input data (endowed with the χ^2 distance) are mapped, through eigenvalue decomposition, into the space of factors F (on points/rows) and G (on projection axes/columns). The factors are the principal axes of inertia, where the mass, a probability measure, is defined from the marginal distributions of the data array. The row and column clouds of points have identical inertia. The factor space decomposition of inertia is ordered by decreasing inertia, λ_α, $\alpha = 1, 2, \ldots$, associated with the factors.

The cloud of observables, I, is denoted $N(I)$. The cloud of attributes, J, is similarly denoted $N(J)$. The projection of observable $i \in I$ on the αth factor is $F_\alpha(i)$. This is the distance of the profile, f_J^i from the mean profile, f_J. Similarly $G_\alpha(j)$ measures the distance, as projected on the axis α, of the profile f_I^j from the mean profile, f_I. See [29, p. 126].

The χ^2 statistic assesses how informative our data are, relative to the baseline case of a data array frequency value, f_{ij} being equal to the product of the marginal frequency distributions. We have the following expression for this decomposition, for rows, i, and columns, j:

$$f_{ij} = f_i f_j \left(1 + \sum_{\alpha=1,\ldots,n} \lambda_\alpha^{-\frac{1}{2}} F_\alpha(i) G_\alpha(j) \right). \tag{6.2}$$

6.8.1   Clouds of Points, Masses and Inertia

We use I to denote the labels or indices of the observation set, and J to denote the labels or indices of the attribute set.

The data to be analysed may be a contingency table, meaning a cross-tabulation of frequency counts; or they can be any array of categorical data, or non-negative continuous-valued data. (The data are to be non-negative, due to the definition of mass, which will now be given. However, what is important is that the row and column marginals, viz. row sums and column sums constituting the row masses and column masses, be positive. Discussion of this, with implications, is in [131, pp. 59–60].)

Here, the given data table to be analysed is denoted $k_{IJ} = \{k_{IJ}(i,j) = k(i,j); \; i \in I, \; j \in J\}$. We have $k(i) = \sum_{j \in J} k(i,j)$; we define $k(j)$ analogously, and $k = \sum_{i \in I, j \in J} k(i,j)$. Next, we convert our data to frequency values. The data table becomes $f_{IJ} = \{f_{ij} = k(i,j)/k; \; i \in I, \; j \in J\} \subset \mathbb{R}_{I \times J}$. Similarly, the marginal distributions are f_I, defined as $\{f_i = k(i)/k; \; i \in I, \; j \in J\} \subset \mathbb{R}_J$, and f_I is defined analogously.

Regarding notation here, $\mathbb{R}_J$ is the real space (i.e. comprising real values), the dimensionality of which is the cardinality of the set J. So the number of values in a point is the cardinality of the set J. $\mathbb{R}_{I \times J}$ is the coupling of real values that have dimension the cardinality of the set I, and real values that have dimension the cardinality of the set J. It is the outer product of the real spaces, $\mathbb{R}_I$ and $\mathbb{R}_J$.

The f functions can be considered (relatively and empirically defined) probabilities. We have, $\sum_i f_i = 1$, $\sum_j f_j = 1$ and $\sum_{i,j} f_{ij} = 1$.

The moment of inertia of a cloud of points in a Euclidean space, with both distances and masses defined, is the sum, for all elements of I, of the products of mass by distance squared from the centre of the cloud:

$$M^2(N_J(I)) = \sum_{i \in I} f_i \| f_J^i - f_J \|_{f_J}^2 = \sum_{i \in I} f_i \rho^2(i). \tag{6.3}$$

In the last equality, ρ is the Euclidean distance from the cloud centre, and f_i is the mass of element i. The mass is defined from the marginal distribution of the input data table.

Returning to the first right-hand-side term in equation (6.3), the conditional distribution of f_J knowing $i \in I$, also termed the jth profile with coordinates indexed by the elements of I, is

$$f_J^i = \{f_j^i = f_{ij}/f_i = (k_{ij}/k)/(k_i/k); \; f_i \neq 0; \; j \in J\},$$

and likewise for f_I^j.

The cloud of points consists of the couple: profile coordinate and mass. We have $N_J(I) = \{(f_J^i, f_i); \; j \in J\} \subset \mathbb{R}_J$, and again similarly for $N_I(J)$. In correspondence analysis there is complete symmetry between these clouds, $N_J(I), N_I(J)$, in the sense that both are mapped into the factor space endowed with the Euclidean distance. The factors can be defined in either of these two spaces. Therefore, for computational reasons, we define the factors on the lower dimensional cloud, and then map the other into the factor space. As noted above, the cloud of row points is in the space $\mathbb{R}_J$; the cloud of column points is in $\mathbb{R}_J$; the dimensionality of the cloud of row points is $|I|$; and the dimensionality of the cloud of column points is $|J|$. Of course that is the case when considering the two clouds separately. Since these spaces, $\mathbb{R}_J$ and $\mathbb{R}_I$ are inherently related, they are referred to as dual spaces; that is, the two spaces are in duality due to the transition formulas relating one to the other. The eigenvectors and eigenvalues are found (sometimes referred to as diagonalization) in the space of lower cardinality, $\min\{|I|, |J|\}$.

From equation (6.3), it can be shown that the cloud inertia is

$$M^2(N_J(I)) = M^2(N_I(J)) = \|f_{IJ} - f_I f_J\|^2_{f_I f_J} = \sum_{i \in I, j \in J} (f_{ij} - f_i f_j)^2 / f_i f_j. \qquad (6.4)$$

The term $\|f_{IJ} - f_I f_J\|^2_{f_I f_J}$ is the χ^2 metric between the probability distribution f_{IJ} and the product of marginal distributions $f_I f_J$, with as centre of the metric the product $f_I f_J$.

In the foregoing, let us show relation (6.4). The contribution of element $i \in I$ to the cloud's inertia is $f_i \| f_J^i - f_J \|^2_{f_J}$ (equation (6.3)). This expresses the following: mass, f_i; and the centred Euclidean distance squared, with centre at f_J. Rewriting the profile, $f_J^i = \frac{f_{iJ}}{f_i}$, in the above expression gives $f_i \sum_j (\frac{1}{f_j}(\frac{f_{ij}}{f_i} - f_j)^2)$. Note how the centred distance requires the $1/f_j$ weighting. Now consider the contributions of all $i \in I$. We have

$$\sum_i \sum_j f_i \left(\frac{1}{f_j} \left(\frac{f_{ij}}{f_i} - f_j \right)^2 \right) = \sum_i \sum_j \frac{f_i}{f_j} \left(\frac{f_{ij} - f_i f_j}{f_i} \right)^2 = \sum_i \sum_j \frac{(f_{ij} - f_i f_j)^2}{f_i f_j}.$$

The foregoing expression is the χ^2 statistic, expressing the sum of observed values, O, relative to expected values, E: $(O - E)^2 / E$. This is to say that the expected characteristic of our data is a product distribution of the mean mass vectors (denoted f_I and f_J). Furthermore, the mean mass vectors, determined from the input data marginal distributions, can be considered as empirical probability distributions.

In the factor space, we have the following. Correspondence analysis produces an ordered sequence of pairs, called factors, (F_α, G_α), associated with real numbers called eigenvalues $0 \leq \lambda_\alpha \leq 1$. The number of eigenvalues and associated factor couples is $\alpha = 1, 2, \ldots, N = \inf(| I | -1, | J | -1)$, where $| \cdot |$ denotes set cardinality. We denote by $F_{\alpha(I)}$ the set of values of the factor of rank α for elements i of I; and similarly by $G_{\alpha(J)}$ the values of the factor of rank α for all elements j of J. We see that F is a function on I, and G is a function on J.

The moment of inertia of the clouds $N_J(I)$ and $N_I(J)$, relative to the α axis, is λ_α. Decomposition of the cloud's inertia is then as follows:

$$M^2(N_J(I)) = \sum_{\alpha=1,\ldots,N} \lambda_\alpha = \sum_{i \in I} f_i \rho^2(i). \qquad (6.5)$$

In greater detail, we have for this decomposition,

$$\lambda_\alpha = \sum_{i \in I} f_i F_\alpha^2(i) \quad \text{and} \quad \rho^2(i) = \sum_{\alpha=1,\ldots,N} F_\alpha^2(i). \qquad (6.6)$$

6.8.2   Relative and Absolute Contributions

Contributions to inertia are fundamental in order to define the mapping into the factor space. Contributions by row points, or by column points, in their respective dual spaces, define the importance of those given data elements for the constructed mapping. Supported by the experimental results to be reported on in the following sections, we will use the average contribution to the inertia as a measure of cloud concentration. The inertia is the fundamental determinant of not only relative positioning, but also essential cloud properties.

The motivation for the use of the contributions to the total inertia of the cloud as the basis for a measure of concentration is as follows. Consider the hypothetical scenario where massive points in the cloud were moved towards the centre or origin, leaving light points to drift away from the centre. Through inertia, we would characterize such a scenario as concentration. Or consider where massive points drift apart, and their inertia contributions outpace the inertia contributions of less massive points that move closer to the origin. Again in that scenario, our inertia measure of concentration would be appropriate for quantifying the lack of concentration. In these hypothetical scenarios, we see how contribution to inertia is a key consideration for us. Inertia is more important than projection (i.e. position) *per se*.

We now look at absolute versus relative contributions to inertia. The former are the more relevant for us. This will be seen in our experiments below. What we consider for the attributes (measurements, dimensions) holds analogously for the observations.

- $f_j \rho^2(j)$ is the absolute contribution of attribute j to the inertia of the cloud, $M^2(N_I(J))$. Therefore, from expressions (6.6), this absolute contribution of point j is also $f_j \sum_{\alpha=1,...,N} F_\alpha^2(j)$.

- $f_j F_\alpha^2(j)$ is the absolute contribution of point j to the moment of inertia λ_α.

- $f_j F_\alpha^2(j)/\lambda_\alpha$ is the relative contribution of point j to the moment of inertia λ_α. We noted in (6.6) that $\lambda_\alpha = \sum_{j \in J} f_j F_\alpha^2(j)$. So the relative contribution of point j to the moment of inertia λ_α is $f_j F_\alpha^2(j)/\sum_{j \in J} f_j F_\alpha^2(j)$. The total relative contribution of j, over all $j \in J$, is 1. The total contribution over all factors, indexed by α, then becomes N here. So the mean contribution (here, the mean relative contribution) of the attributes, expressed as a percentage, is $100N/|J|$.

We now have the technical machinery needed to evaluate data clouds in very high dimensions. We will keep our cloud of observables, $N(I)$, small. It is in a $|J|$-dimensional space. That dimensionality, $|J|$, will be very large. That is to say, the cloud of what we take as attributes, $N(J)$, will be huge. While the cloud itself, $N(J)$, is huge, each point in that cloud, $j \in J$, is in a space of dimension $|I|$, which is not large.

Now we will carry out our evaluations. Our choice of cloud cardinality and dimensionality are motivated by inter-study comparison.

6.9 Evaluation 1: Uniformly Distributed Data Cloud Points

Uniformly distributed values in $[0, 1]$ were used for five data clouds, each of 86 points, in dimensionalities of 100, 1000, 10,000, 100,000 and 1,000,000. In the usual analysis perspective, we have 86 observations, and the dimensionalities are associated with the attributes or features. These input data are therefore dense in value. The results obtained are shown in Table 6.2.

Note how increasing dimensionality implies the following. We base concentration, or compactness, on the absolute contribution to the inertia of the factors. The average absolute contribution to the factors tends towards zero. The standard deviation also approaches zero. Thus the cloud becomes more compact.

We provide the median as well as the mean as an indication of distributional characteristics of the absolute contribution that we are examining. We observe a relatively close match between mean and median values, implying an approximate Gaussian distribution of the absolute contributions. For all cases (including the 1,000,000-dimensional case), we checked that the distributions of absolute and relative contributions, and norms squared of the input data, are visually close to Gaussian.

The maximum projection values, which do not decrease, serve to show that concentration

Contributions to inertia of factors by the columns				
Dim.	Contributions	Mean	Std. dev.	Median
100	Absolute	0.01322144	0.0005623589	0.01325343
	Relative	0.86	0.04588791	0.869127
1000	Absolute	0.001331763	5.440168e-05	0.001333466
	Relative	0.086	0.009729907	0.08547353
10000	Absolute	0.0001332053	5.279421e-06	0.0001332981
	Relative	0.0086	0.0009742588	0.008577748
100000	Absolute	1.330499e-05	5.269165e-07	1.332146e-05
	Relative	0.00086	9.783086e-05	0.0008574684
1000000	Absolute	1.330706e-06	5.278487e-08	1.332186e-06
	Relative	8.6e-05	9.788593e-06	8.576992e-05

Maximum factor projection	
Dim.	Projection
100	0.3590788
1000	0.2777193
10000	0.2799913
100000	0.3678137
1000000	0.3750852

TABLE 6.2: Five data clouds, each of 86 points, in spaces of dimensionality 100, 1000, 10,000, 100,000 and 1,000,000. The original coordinate values are randomly uniform in $[0, 1]$.

with increasing dimensionality is a phenomenon relating to the whole cloud, and therefore to the average (or median).

6.9.1   Computation Time Requirements

The largest, uniformly random generated, data set used was of dimensions $86 \times 1,000,000$. In order to create this data array (see the annex to this chapter), an elapsed time of 82.8 seconds was required. Carrying out the main processing, furnishing the results in Table 6.2, involved a basic correspondence analysis of this input data matrix. The projections and contributions (to inertia) of the 86 points were to be determined.

As described in the annex to this chapter, standard processing proved satisfactory for these evaluations. For this large data set, our main processing took an elapsed time of 95.6 seconds. The machine used was a Mac Air, with a 2 GHz processor, and 8 GB of memory, running OS X version 10.9.4. The version of R in use was 2.15.2.

6.10   Evaluation 2: Time Series of Financial Futures

In [175], we use financial futures from the DAX exchange, from around March 2007, denominated in euros. Our data stream, at the millisecond rate, comprised 382,860 records. Each record includes five bid and five asking prices, together with bid and asking sizes in all cases, and action.

Dim.	Contribution	Mean	Std. dev.	Median
100	Absolute	0.01	9.260615e-08	0.01000002
	Relative	0.86	0.05399462	0.8672608
1000	Absolute	0.001	3.297399e-08	0.001000008
	Relative	0.086	0.0121773	0.08518253
10000	Absolute	0.0001000001	2.168381e-08	9.999872e-05
	Relative	0.0086	0.001159708	0.008477465

Maximum factor projection	
Dim.	Projection
100	0.0001054615
1000	0.0002979516
10000	0.0008869227

TABLE 6.3: Embeddings, of dimensionalities 100, 1000 and 10,000, for a financial time series.

We used the sliding window approach to embed the financial signal in spaces of varying dimensionality. Various examples in [175] showed how there may be no "curse of dimensionality", in Belman's [20] famous phrase, in very high-dimensional spaces. There is no such obstacle if we seek out, and make use of, the "remarkable simplicity" [175] of very high-dimensional data clouds.

We extracted one symbol (commodity) with 95,011 single bid values, on which we now report results. These values were continuous and avoided missing values. The data values were integers between 6788 and 6859.5, in steps of 0.5. Very often this signal contained short sequences of successive identical values.

Similar to [175], we define embeddings of this financial signal as follows. Each embedding begins at the following time steps in the financial signal: 1, 1000, 2000, ..., 85,000. The lengths of the success embeddings were, in our three case studies: 100, 1000, 10,000. That provided matrices, in these three case studies, of sizes: $86 \times 100, 86 \times 1000, 86 \times 10,000$. The annex to this chapter provides the R code used.

The results obtained are presented in Table 6.3. The histograms of projections on the factors were visually observed to be Gaussian distributed. We observe how the mean absolute contribution and the median absolute contribution decrease as the embedding dimensionality increases. The standard deviations of absolute and relative contributions decrease too, indicating the increasing concentration. Our measure of concentration is the average (or median) contribution by the embedding dimensionality values (what we may consider as attributes or characterizing features of the "sliding window" over the signal) to the inertia of the factors. We observe also how the maximum projection on the factors does not decrease. This just means that the cloud in the overall sense, and on the whole, gets increasingly compact or concentrated, as the attribute dimensionality increases.

6.11 Evaluation 3: Chemistry Data, Power Law Distributed

6.11.1 Data and Determining Power Law Properties

In our earlier work in [186], we used a set of 1,219,553 chemical structures coded through 1052 presence/absence values, using the Digital Chemistry BCI 1052 dictionary of chemical fragments. That binary-valued matrix was sparse: occupancy (i.e. presence = 1 values) of the chemicals crossed by attribute values was 8.6%.

Our motivation here is to investigate the effect of greatly increasing the attribute dimension. In the next subsection we will develop a novel way to do this. In this subsection we determine the relevant statistical properties of our data.

Here, we will use 425 chemicals from this set, in 1052-dimensional space. We took 425 chemicals in order to have a limited set, $|I| = 425$, in the attribute space, J. Each chemical had therefore presence/absence (i.e. 1 or 0, respectively) values on $|J| = 1052$ attributes. The occupancy of the 425×1052 data set used was 5.9%. Since we wanted this sample of 425 of the chemicals to be representative of the larger set from which they came, we now look at the most important distributional properties.

The marginal distribution, shown in Figure 6.6, is not unlike the marginal distribution displayed in [186]. In that previous work, we found the power law distribution of the chemical attributes to be of exponent -1.23. Let us look at the power law of the baseline distribution function used here, that is, relating to the 425 chemicals.

A power law (see [159]) is a frequency of occurrence distribution of the general form $x^{-\alpha}$, where constant $\alpha > 0$; whereas an exponential law is of the form e^{-x}. For a power law, the probability that a value, following the distribution, is greater than a fixed value is $P(x > x_0) \sim cx^{-\alpha}, c, \alpha > 0$. A power law has heavier tails than an exponential distribution. In practice, $0 < \alpha \leq 3$. For such values, x has infinite (i.e. arbitrarily large) variance; and if $\alpha \leq 1$ then the mean of x is infinite. The density function of a power law is $f(x) = \alpha cx^{-\alpha-1}$, and so $\ln f(x) = -\alpha \ln x + C$, where C is a constant offset. Hence a log–log plot shows a power law as linear. Power laws have been of great importance for modelling networks and other complex data sets.

Figure 6.7 shows a log-log plot based on the 1052 presence/absence attributes, using the 425 chemicals. In a very similar way to the power law properties of large networks (or file sizes, etc.) we find an approximately linear regime, ending (at the lower right) in a large fan-out region. The slope of the linear region characterizes the power law. For this data set, we find that the probability of having more than n chemicals per attribute to be approximately $c/n^{1.49}$ for large n.

The histogram of attributes per chemical, on the other hand, is approximately Gaussian. This is as observed in [186].

6.11.2 Randomly Generating Power Law Distributed Data in Varying Embedding Dimensions

In Section 6.9 we used dense uniformly distributed data. In Section 6.10 our financial futures were slow-moving, in the sense of small variation between successive values. But there too the data were dense and real-valued. Our chemistry context is sparse and boolean-valued (for presence/absence). We use this context to generate data that keep the property of the attributes (i.e. the columns or dimensions) following a power law in regard to their distribution.

To generate new random data sets that fully respect the distributional characteristics of

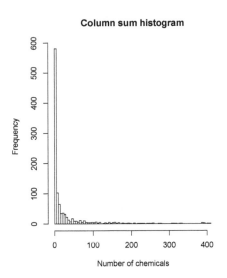

FIGURE 6.6: Histogram of column (i.e. chemical attribute) sums.

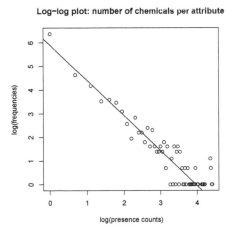

FIGURE 6.7: Log–log plot of numbers of chemicals per attribute, based on the data set of 425 chemicals.

425 chemicals	
Dim.	Exponent
1052	−1.49
10520	−1.75
105200	−1.64
1052000	−1.78

TABLE 6.4: Power law exponents for generated chemical data, with 425 chemicals, with presence/absence (respectively 1 or 0) in attribute dimensions 1052, 10,520, 105,200 and 1,025,000.

our known data, we will use the distribution function that is displayed in Figure 6.7. This is the data distribution of coding attributes that characterize the chemicals (i.e. presence of molecules).

In line with our earlier notation, the marginal distribution in Figure 6.7 is f_J for attribute set, J. The chemicals set is I. The presence/absence cross-tabulation of chemicals by their attributes is, in frequency terms, f_{IJ}. The (i, j) elements, again in frequency terms, are $f_{i,j}$. In whole-number terms, representing presence or absence, 1 or 0, the chemicals–attributes cross-tabulation is denoted k_{IJ}.

We generate a new data set that cross-tabulates a generated set of chemicals, I', crossed by a generated set of attributes, J'. Let $|\cdot|$ denote cardinality. We randomly sample (uniformly) $|J'|$ values from k_J. Therefore we are constructing a new, generated set of attribute marginal sums. The generated values are of the same distribution function. That is, both $f_{J'} \sim f_J$ and $k_{J'} \sim k_J$. The next step is to consider the newly generated chemicals, in the set I', of cardinality $|I'|$. Given $k_{j'}$, we generate $|k_{j'}|$ values of 1 in the set of $|I'|$ elements. In this way, we generate the chemicals that contribute the $k_{j'}$ attribute presences found for attribute j'.

For the generated chemical data, we use 425 chemicals, in attribute spaces of dimensions 1052, and then 10 times this, 100 times this, and 1000 times this dimensionality.

See the annex to the chapter for the R code used (under "Evaluation 3"). This code shows the case of 1000 times the dimensionality. That is, for 425 chemicals with 1052 presence/absence or 1/0 values, we generate a matrix of 425 chemicals × 1,052,000 presence/absence attributes. For the 425 × 1052 matrix, we have 26,405 presence values, and a density (i.e. presence or 1 values) of 5.9%. For the generated 425 × 1,052,000 presence/absence attributes, we have 5,645,075 presence values, and a density of 1.26%.

Figure 6.8 displays the marginal distributions. This shows visually how well our generated data approximate the original data. Let us also look at how close the power law distributional properties are. Table 6.4 lists the power law exponents for our generated data sets.

Table 6.5 shows clearly how the absolute contribution to the inertia of the factors, which is mass times distance squared, becomes of smaller mean value, and of smaller standard deviation (hence the mean is a tighter estimate), as dimensionality increases. The degree of decrease of the mean value is approximately linear in the increase of dimensionality (i.e. tenfold for each row of Table 6.5). Once again, we show very conclusively how increasing dimensionality brings about a very pronounced concentration of the data cloud that is considered. As dimensionality increases, the cloud becomes more more compact, that is, much more concentrated.

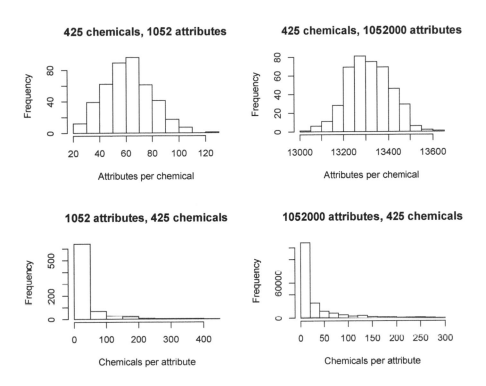

FIGURE 6.8: Histograms of marginal distributions of the original 425×1052 chemicals by attributes, and the generated data with similar marginal distributions, of $425 \times 1,052,000$ chemicals by attributes. Marginal distribution values greater than 0 were taken into account.

425 chemicals	Absolute contribution		Max. projection
Dimensionality	Mean	Std. dev.	
1052	0.01161321	0.007522956	16.27975
10520	0.00133034	0.002798697	12.31945
105200	0.000140571	0.0002946923	10.91465
1052000	1.39319e-05	2.919471e-05	11.06306

TABLE 6.5: Four hundred and twenty-five chemicals with presence/absence values on the following numbers of characterizing attributes: 1,052, 10,520, 105,200 and 1,052,000. The dimensionality of the space in which the chemicals are located is given by the number of characterizing attributes.

6.12 Application 1: Quantifying Effectiveness through Aggregate Outcome

We are concerned with quantifying effectiveness, or aggregate outcome, of a human or machine action. In this first application, we look at the computational requirements of the processing.

Traditionally the "impact" of an action is considered in computational terms as what happens when an algorithm is executed. In general, an algorithm is a chain of processing actions. Now consider some specified action, with some desired or targeted outcome. In order to express and model (mathematically, computationally) general human or social, or other, scenarios, we will relax what we take as "impact", in favour of "effectiveness". We define effectiveness as the general, and hence aggregate, outcome. In the space of all actions, our initial action will be a point. Then all the actions considered are points in the space of actions. Together, they comprise a cloud of points. Finally, the aggregate outcome of all these actions is the centroid (mean, centre of gravity) of the cloud of actions.

In [192], we take "instigational" actions and examine them relative to an aggregate outcome. The latter is an average profile. In that work, the actions were tweets (so-called "micro-blogs"), in a Twitter context. We were studying the process of communicative action, with Jürgen Habermas's social-political theory of communicative action as motivation for that work. We used successive Twitter campaigns (relating to environmental citizenship). We wanted to see how well an initiating tweet would be matched against an overall campaign average. We used a semantic embedding of all of our tweets that were studied. This semantic embedding was based on the textual content of the tweets. Correspondence analysis provides such a latent semantic embedding.

Consider, for our purposes here, an overall or global average. From Section 6.8 we know that a profile element, f_J^i (e.g. tweet profile indexed by i), is with reference to the centre f_J (which is the marginal probability distribution function, i.e. empirically defined from the frequencies, of the textual content attributes used).

For such categorical data, as is the case for us in this application, let us write $\mathbb{Z}_+^{|J|}$ for the space of $|J|$-dimensional, non-negative integers. We have that $k_{iJ}, k_J \in \mathbb{Z}_+^{|J|}$. Recall that we use k for our initially given data, and we use f for these data when converted to frequencies. If presences and absences are at issue, then we will have $k_{iJ}, k_J \in \{0, 1\}^{|J|}$.

In the factor space, the norm squared of tweet i is $\sum_\alpha F_\alpha(i)^2$. This is equal to the χ^2 distance on profiles: $\|f_J^i - f_J\|_{f_J}^2$. We have discussed these terms in Section 6.8.1. From the practical point of view, what we are showing here are the operations to be performed on the input data, in order to determine key measures in the factor space. Knowing these identities, for very big data we are able to determine all that we need in our analytics, without the need for an eigenreduction, or for any other, computationally demanding, processing. We further elaborate on this conclusion in the following subsection.

6.12.1 Computational Requirements, from Original Space and Factor Space Identities

To carry out validation of what was concluded in the previous section, we used a matrix of uniformly distributed values (in the unit open interval, $(0, 1)$), comprising a matrix of dimensions 126×2000. The 126 points in the 2000-dimensional space comprise the cloud in the space $\mathbb{R}_J$. In the 126-dimensional space, the 2000 points comprise the cloud in the dual space, $\mathbb{R}_I$.

From Section 6.8.1, we have distance invariance (with no loss of generality, we use squared distances here). In terms of original values, k, the derived frequency terms, f, and the factor coordinates, F, we have the identities

$$\sum_j \frac{k}{k_j} \left(\frac{k_{ij}}{k_i} - \frac{k_{i'j}}{k_{i'}} \right)^2 = \sum_j \frac{1}{f_j} \left(\frac{f_{ij}}{f_i} - \frac{f_{i'j}}{f_{i'}} \right)^2 = \sum_\alpha (F_\alpha(i) - F_\alpha(i'))^2. \qquad (6.7)$$

Turning attention to the squared distance to the origin (i.e. the norm squared), for a given point, we have

$$\sum_j \frac{k}{k_j} \left(\frac{k_{ij}}{k_i} - \frac{k_j}{k} \right)^2 = \sum_j \frac{1}{f_j} \left(\frac{f_{ij}}{f_i} - f_j \right)^2 = \sum_\alpha (F_\alpha(i) - 0)^2 = \sum_\alpha F_\alpha(i)^2. \qquad (6.8)$$

For a given population of outcome acts (i.e. outcome units, such as actions or events), the computational requirements are as follows. Our ith causal or instigational action, measured in our parameter space, is $\{k_{ij} | j \in J\}$. Our aggregate outcome is the vector, $f_J = \{f_j = \frac{k_j}{k} | j \in J\}$. Then our effectiveness measure is $\sum_j \frac{k}{k_j} \left(\frac{k_{ij}}{k_i} - \frac{k_j}{k} \right)^2$. Having $n = |I|$ and $m = |J|$ means that k is of computational complexity $O(nm)$; f_J, given k, is $O(m)$; and our effectiveness measure, given the component values, is $O(m)$. Hence the overall computational complexity is linear in the data we have, that is, $O(nm)$. This dispenses with the $O(n^3)$ (for $n < m$) computational requirement of eigenreduction (i.e. determining the factor space mapping).

In conclusion, for this approach to quantifying effectiveness, that is, the aggregate outcome, from human or machine action, we require computational time that is linear in the number of actions, and in the number of attributes used to characterize the actions.

6.13 Application 2: Data Piling as Seriation of Dual Space

In [169] it was noted how very high-dimensional data become ultrametric. That is to say, a tree structure (not unique, however) can well summarize such a very high-dimensional cloud of points. In [53] it was shown how hierarchical topologies can be perfectly scaled in one dimension. In this section these results are exploited, from a practical point of view. These outcomes will also be related to the use of such a scaled hierarchical or ultrametric topology as the starting point for endowing a Baire hierarchical clustering on the data cloud.

Based on the data preparation, and the observed data piling, we have the following. From equation (6.5), our average λ_α tends to 0 with increasing dimensionality. Hence, from equation (6.3), our data, in frequency terms, are such that $f_{ij} \to f_i f_j$ for all i, j. Otherwise written, we have that $f_{IJ} \to f_I f_J$ for observables set I and attributes set J.

In summary, we have (1) concentration with dimensionality on the average profile; (2) yet, as indicated in Tables 6.2–6.5, differentiation that remains and is manifested in our data through the projections (or loadings) on the factors. The latter is also the Chebyshev or L_∞ distance.

From equation (6.2), such concentration or "piling" tends towards $f_{ij} = f_i f_j$ for all $i \in I$, $j \in J$. In such a case, a zero value is given by the χ^2 statistic. The vector f_J is mapped onto the origin in the factor space, endowed with the Euclidean distance. It is therefore mapped onto the vector 0. In the original space, endowed with the χ^2 metric, f_J is of course not zero-valued. (As is clear, it constitutes the column masses; it is defined as the column sums, divided by the overall sum of the $I \times J$ matrix.) Consequently, under these circumstances of extreme piling, we can write our row vectors as $f_i f_J = \{f_i f_j | 1 \leq j \leq |J|\}$. Thus, row i has coordinate f_i on the vector f_J.

The effect of the piling is to condense the relationships in the set of row vectors f_I to just their projections on the axis, f_J. We can write that as $|J| \to \infty$, so $f_{iJ} \to f_i$. Clearly, this is a particular projection. We can consider f_{iJ} being mapped onto any other axis as well.

Using the sparse $10,317 \times 34,352$ (0.2854% non-zero values) data matrix from [185], we took (1) one (uniform) random projection of the 10,317 rows; (2) the mean of 99 random projections, as discussed in [185]; and (3) f_I, the row masses, as defined on this data set. The latter, f_I, is the (empirical) distribution function of the set of documents that comprised the row set, I. The correlation of the distribution function, f_I, with the one random projection, was 0.9905881. The correlation of the distribution function, f_I, with the mean of the 99 random projections was 0.9905582.

A question arises as to why we would want to use random projections, rather than just take the ordered f_I values, as the starting point for inducing the Baire hierarchical clustering tree. We use random projection because it differentiates vectors that are different but which have the same marginal sum values, and thus $f_{iJ} \neq f_{i'J}$ and $f_i = f_{i'}$. We may ask a further question: could we have $f_{iJ} \neq f_{i'J}$ such that the random projections of these two vectors are identical? Our answer is that this cannot be ruled out. However, through the randomization, its probability of occurrence is very small. Taking the mean of random projections further diminishes the probability of such a hash collision.

6.14   Brief Concluding Summary

We have explored a number of aspects of metric scaling of very high-dimensional data. This metric scaling is the mapping of our data into a factor (or latent semantic) space, endowed with the Euclidean distance. Our input data can be qualitative or categorical, quantitative and non-negative, or a mixture of all of these.

We explored a wide range of evaluation settings. We then reviewed applications of this work. We have shown that is is easy and straightforward to analyse data that are in very high attribute dimensions (or feature dimensions, in other words, typically the number of columns of our input data matrix). Of course one needs to understand the nature of one's analysis. It is not a "black box" process. Instead it is necessary to investigate how to "let the data speak".

Future work is likely to see even further development and wider application of issues discussed in Sections 6.12 and 6.13.

6.15   Annex: R Software Used in Simulations and Evaluations

The following code is accessible on the book's website, together with data sets used here. For convenience, given both the data simulations at issue here, and the related evaluations carried out on these simulations, this software is listed here.

6.15.1 Evaluation 1: Dense, Uniformly Distributed Data

```
set.seed(4781)
x <- runif(86000000)
z100 <- buildmat(x, 86, 100)    # See below for the function, buildmat
z1000 <- buildmat(x, 86, 1000)
z10000 <- buildmat(x, 86, 10000)
z100000 <- buildmat(x, 86, 100000)
z1000000 <- buildmat(x, 86, 1000000)

xtab <- t(z100)  # Transpose used for efficiency (eigen-reduction).
# PROCESSING (SEE BELOW)
xtab <- t(z1000)
# PROCESSING
xtab <- t(z10000)
# PROCESSING
xtab <- t(z100000)
# PROCESSING
xtab <- t(z1000000)
# PROCESSING

buildmat <- function(sig, numsegs, lenseg) {
    # Input signal, number of segments, segment length.
    # From a signal, take successive segments of specified length,
    # assemble row-wise in a matrix.
    outmat <- sig[1:lenseg]    # Very first segment.
    finseg <- numsegs - 1    # For final segment.
    for (i in 1: finseg) {      # For 2nd, 3rd, ... final segment..
        loind <- (i * lenseg) + 1
        hiind <- (i * lenseg) + lenseg
        nextseg <- sig[loind : hiind]
        outmat <- rbind(outmat, nextseg)  # Build up by row.
    }
    outmat
}


# PROCESSING

# THE PROCESSING: CORRESPONDENCE ANALYSIS (EIGEN-REDUCTION, USING eigen)
tot <- sum(xtab); fIJ <- xtab/tot; fI <- apply(fIJ, 1, sum); fJ <- apply(fIJ, 2, sum)
if (length(fI[fI <= 0]) >= 1) cat("Note and check: fI terms le 0. <-- 1.\n")
if (length(fJ[fJ <= 0]) >= 1) cat("Note and check: fJ terms le 0. <-- 1.\n")
fI[fI <= 0] <- 1; fJ[fJ <= 0] <- 1; fJsupI <- sweep(fIJ, 1, fI, FUN=''/'')
fIsupJ <- sweep(fIJ, 2, fJ, FUN="/"); s <- as.matrix(t(fJsupI))
s1 <- sweep(s, 1, sqrt(fJ), FUN="/"); s2 <- sweep(s1, 2, sqrt(fJ), FUN="/")
sres <- eigen(s2,symmetric=T)
# Eigenvectors divided row-wise by sqrt(fJ):
evectors <- sweep(sres$vectors, 1, sqrt(fJ), FUN="/")


# PROJECTIONS ON FACTORS OF ROWS AND COLUMNS
```

```
rproj <- as.matrix(fJsupI) %*% evectors  #This is v large, hence bypass.
temp  <- as.matrix(s2) %*% sres$vectors
cproj <- sweep(sweep(temp,1,sqrt(fJ),FUN="/"),2,sqrt(sres$values),FUN="/")


# CONTRIBUTIONS TO FACTORS BY ROWS AND COLUMNS
# Contributions: mass times projection distance squared.
temp <- sweep( rproj^2, 1, fI, FUN="*")
# Normalize such that sum of contributions for a factor equals 1.
sumCtrF <- apply(temp, 2, sum)
# Note: Obs. x factors. Read cntrs. with factors 1,2,... from cols. 2,3,...
rcntr <- sweep(temp, 2, sumCtrF, FUN="/")


# ABSOLUTE CONTRIBUTIONS (of very large row set), THEN REL CONTRIBUTIONS, CTR
cat("Abs. contr. mean, sd, med.; rel. contr. - CTR: mean, sd, med.; Max rproj valu
cat("Abs: ", mean(apply(temp,1,sum)), sd(apply(temp,1,sum)), median(apply(temp,1,s
"Rel: ", mean(apply(rcntr,1,sum)), sd(apply(rcntr,1,sum)), median(apply(rcntr,1,su
"Maxproj: ", max(rproj))
```

6.15.2   Evaluation 2: Financial Futures

```
# For our financial data, function buildmat defines as follows
# loind, hiind, i.e. the indexes of the segments.
buildmat <- function(sig, numsegs, lenseg) {
    # Input signal, number of segments, segment length.
    # From a signal, take successive segments of specified length,
    # assemble row-wise in a matrix.
    outmat <- sig[1:lenseg]      # Very first segment.
    finseg <- numsegs - 1     # For final segment.
    for (i in 1: finseg) {        # For 2nd, 3rd, ... final segment..
        loind <- (i * 1000) + 1          # For our financial data,
        hiind <- (i * 1000) + lenseg     # we alter loind, hiind.
        nextseg <- sig[loind : hiind]
        outmat <- rbind(outmat, nextseg)  # Build up by row.
    }
    outmat
}


x <- scan("ob5x.csv")    # Read 95011 values of financial futures.
x100 <- buildmat(x, 86, 100)
x1000 <- buildmat(x, 86, 1000)
x10000 <- buildmat(x, 86, 10000)
# We now have arrays of dimensions, 86 x 100, 1000, 10000


xtab <- t(x100)
# PROCESSING (SEE ABOVE)
xtab <- t(x1000)
# PROCESSING
xtab <- t(x10000)
# PROCESSING
```

6.15.3 Evaluation 3: Chemicals of Specified Marginal Distribution

```
x <- read.table("inputdata.txt")    # 425 x 1052 matrix.
100*sum(x)/(nrow(x)*ncol(x))        # Occupancy found to be 5.9%
hist(apply(x,2,sum), xlab="Number of chemicals", ylab="Frequency",
  nclass=100,main="Column sum histogram")
# Now investigate marginal law of the chemicals.
comptes <- hist(apply(x,2,sum),nclass=100)$counts
comptes[comptes == 0] <- 1  # Done to allow flexible use of log.
length(comptes)  # We find 82 counts or histogram bins.


plot(log(1:82), log(comptes), xlab="log(presence counts)", ylab="log(frequencies)")
title("Loglog plot: number of chemicals per attribute")
# Once beyond 24th bin, there are many 0s. This gives rise to 0 values in the plot.
# So we will just fit the regression line to the first 24 bins.
abline(lsfit(log(1:24), log(comptes[1:24])))
lsfit(log(1:24), log(comptes[1:24]))$coefficients  # Provides the slope.


# Just to observe that the marginal distribution, or masses, of the
# chemicals are approximately Gaussian.
hist(apply(x,1,sum),nclass=20)
qqnorm(apply(x,1,sum))


# Now for our long-tailed attributes, the marginal distribution is as follows.
xJ <- apply(x, 2, sum)


# Here is how we generate data with the same marginal distribution.
newnr <- nrow(x); newnc <- 1000*ncol(x)    # New dimensions.
xnew <- matrix( rep(0,newnr*newnc), nrow=newnr, ncol=newnc) # Initialize to 0.
set.seed(4781)              # For reproducibility, set seed.
randomcols <- floor( 1+ (ncol(x)-1) * runif(floor(newnc/4)) )
# In the foregoing, we generate newnc new column numbers; \in [1,ncol(x)]
for (j in 1:length(randomcols)) { # For each col. (The cols. are power law distributed.
   xj <- xJ[randomcols[j]]  # The marginal sum of that col.
   randomrows <- floor(1+(newnr-1)*runif(xj))  # Generate nrow(x) row nos. randomly.
   xnew[randomrows,j] <- 1
}  # End.
x1000nc <- xnew  # New dimensions: 425 x 1052000
```

Part IV

New Frontiers: New Vistas on Information, Cognition and the Human Mind

7

On Ultrametric Algorithmic Information

7.1 Introduction to Information Measures

How best to quantify the information of an object, whether natural or artefact, is a problem of wide interest. A related problem is the computability of the object. We describe practical examples of a new way to address this problem. By giving an appropriate representation to our objects, based on a hierarchical coding of information, we exemplify how it is remarkably easy to compute complex objects. Most interestingly, the algorithmic complexity is related to the length of the class of objects, rather than to the length of the object.

Brooks [39] asserted a great challenge for contemporary computer science and information theory: "Shannon and Weaver performed an inestimable service by giving us a definition of information and a metric for information as communicated from place to place. We have no theory however that gives us a metric for the information embodied in structure ... this is the most fundamental gap in the theoretical underpinning of information and computer science."

The notion of ultrametric information was introduced by [115], both to handle interactive as opposed to static information, and by taking a dynamic view of information, with analogies to metric or Kolmogorov–Sinai information. Here we pursue a view of algorithmic or computational information, which is extended to account for an ultrametric embedding of the object that is computed.

Shannon information is oriented towards communication. While Shannon information is based on the freedom of choice that is possible when transmitting a message, Kolmogorov information, or algorithmic information, is a measure of the information content of individual objects. The Kolmogorov complexity of a string is the size of the shortest program in bits that computes the string. It is concerned therefore with strings, and moreover (finite or infinite) binary strings. An object, expressed as a binary string, has complexity which is its shortest string description, because this also defines the shortest program, or decision tree, to compute it. The shortest effective description length has become known as Kolmogorov complexity, even if precedence may be due to Solomonoff [140, p. 90]. From Solomonoff's work on the "algorithmic theory of descriptions" has come the minimum description length principle as a computable and practical information measure [18, 241, 211]

We approach this problem of expressing information and computability, relating to complexity and generation, respectively, in an innovative way. A key role is played by representation (i.e. object or data encoding). We need to consider carefully the data description related to the observing of the object; and the display associated with the data description. These two issues amount to the mapping of the object to data, and data to object, respectively. There is enormous latitude for representation. We must choose expeditiously, based on our objectives, which may include interpretation of the data or the event or phenomenon.

This work builds on [169, 171] in the following ways. Such work points to the crucial role played by data encoding, or representation, for many purposes (including search and display). In [169] it is shown how *if* we have an ultrametric embedding of our data – otherwise expressed, a hierarchical or tree structuring of our data – then it is possible for search operations to be carried out in constant, or $O(1)$, time. In this chapter, we also presuppose a given ultrametric embedding (or hierarchical structuring) of our data. In general terms we are dealing with n objects characterized by m attributes. Classically, a complete description of an object by means of its attributes leads to an expression for the object's complexity that is defined from the set of its m attributes. Given the ultrametric embedding, we look instead at the object's complexity in terms that are relative to the population of n objects. If the hierarchy is a meaningful one (e.g. expressing biological reproduction), then we have a very new perspective on the computability of an object.

Section 7.2 provides background on an important tool used in subsequent sections, the Haar wavelet transform carried out *on* a hierarchy. It allows us to go well beyond a hierarchy as just a display device or visualization, and instead to carry out operations on the hierarchy, expressing operations in an ultrametric space. We set the scene for later parts of this chapter through a discussion of the stepwise approximation scheme that we can establish, for various objects, and that defines the Haar wavelet transform, in this case, of a dendrogram. "Dendrogram" is the term used for the particular tree, discussed in the next section, that is induced on, or determined from, object/attribute data. In this chapter, the use of the term "hierarchy" is always as a synonym for these terms.

In Section 7.4 we consider a hugely simplified face recognition case study. Once we presuppose a representation or encoding of a face, then any given face is generated by very simple calculations on faces. We link this work with some recent directions of study in the psychology literature of human recognition behaviour.

In dealing with faces, we have carefully selected a case study to exemplify a new approach to computability, in the sense of generation of an object and, related to this, the inherent complexity of an object.

In summarizing and concluding, Sections 7.5 and 7.6 provide further general discussion.

7.2 Wavelet Transform of a Set of Points Endowed with an Ultrametric

While the discussion in this section is fully self-contained, mention may be made of the related discussion in Section 3.6.3. Different agglomerative hierarchical classifications are used.

A wavelet transform is a decomposition of an object, typically an image or signal, into an ordered set of *detail* "versions" of the data, and an overall *smooth* [223]. From the details, with the smooth, the data can be exactly reconstructed. In the case of the Haar wavelet transform, the details and the smooth are defined from differences and sums, respectively. We will see how this works using a concrete example.

The wavelet transform has been extended to ultrametric topologies by, for example, [120, 121]. The wavelet transform has traditionally been used for image and signal processing, based on functions in Hilbert space. In [173] we showed, with a wide range of examples and case studies, how this transform can be easily implemented on tree structured data. Without loss of generality, we assume that the tree is binary, rank-ordered, rooted, and, for practical application, labelled. Such a tree is often referred to as a dendrogram. The tree distance is an ultrametric and, reciprocally, we endow a data set with an ultrametric by structuring it as a tree.

As a small data set consider the first eight observations in the very widely used Fisher iris data [76]. Fisher used this data set, taken from [5], to introduce the discriminant analysis

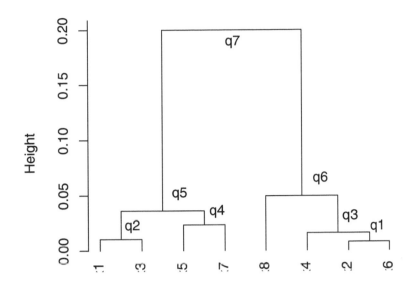

FIGURE 7.1: Minimum-variance or Ward hierarchy of the data shown in Table 7.2. The clusters are labelled q_1, q_2, etc.

method that bears his name. By range-normalizing (i.e. subtracting the minimum value of each variable, and dividing by the range) in Table 7.1, we obtain Table 7.2.

The minimum-variance or Ward agglomerative clustering hierarchy was built (with constant weights on the observations), and is shown in Figure 7.1. The minimum-variance agglomeration criterion, with Euclidean distance, is used to induce the hierarchy on the given data. We could use some other agglomerative criterion. However, the minimum-variance one leads to more balanced dendrograms [166, 171], with knock-on implications for computational requirements for average time tree traversal.

From input Table 7.2 and the dendrogram of Figure 7.1, we carry out the wavelet transform. The transform is shown in Table 7.3, and is also displayed in Figure 7.2.

Note that in Table 7.3 it is entirely appropriate that at more smooth levels (i.e. as we proceed through levels $d_1, d_2, \ldots, d_6, d_7$) the values become more "fractionated" (i.e. there are more values after the decimal point). Each detail signal is of dimension $m = 4$, where m is the same dimensionality as the given, input, data. The smooth signal is of dimensionality m also. The number of detail or wavelet signal levels is given by the number of levels in the labelled, ranked hierarchy (i.e. $n-1$): see the columns in Table 7.3 labelled, for details, $d_7, d_6, \ldots, d_1$.

In the following example, vectors x_1 and x_3 are rows 1 and 3 from input Table 7.2. The node q_2 is defined as $q_2 = (x_1 + x_3)/2$, so that $+d_2$ is defined so that $x_1 = q_2 + d_2$ and $x_3 = q_2 - d_2$. In these two expressions for x_1 and x_3, substituting for q_2 yields $+d_2 = x_1/2 - x_3/2$ and $-d_2 = x_3/2 - x_1/2$. In this way, through a node being the equi-weighted average of its child nodes, details and the overall smooth are defined.

To summarize, we begin typically with an object set, each object having values on an attribute set. From this, a hierarchy of the objects is created. Then this hierarchy is further processed. We get a set of details and a smooth vector, such that they suffice for

TABLE 7.1: First eight observations of Fisher's iris data. L and W refer to length and width.

	Sepal.L	Sepal.W	Petal.L	Petal.W
1	5.1	3.5	1.4	0.2
2	4.9	3.0	1.4	0.2
3	4.7	3.2	1.3	0.2
4	4.6	3.1	1.5	0.2
5	5.0	3.6	1.4	0.2
6	5.4	3.9	1.7	0.4
7	4.6	3.4	1.4	0.3
8	5.0	3.4	1.5	0.2

TABLE 7.2: First eight observations of Fisher's iris data. L and W refer to length and width. Values are range-normalized (in each column: minimum subtracted, and divided by range).

	Sepal.L	Sepal.W	Petal.L	Petal.W
1	0.625	0.5556	0.25	0.0
2	0.275	0.0	0.25	0.0
3	0.125	0.2222	0.0	0.0
4	0.0	0.1111	0.5	0.0
5	0.5	0.6667	0.25	0.0
6	1.0	1.0	1.0	1.0
7	0.0	0.4444	0.25	0.5
8	0.5	0.4444	0.5	0.0

TABLE 7.3: The hierarchical Haar wavelet transform resulting from the hierarchy of Figure 7.1, built on the data of Table 7.2. Last data *smooth*, s_7; levels of *detail* from top to bottom (presented left to right), $d_7, d_6, \ldots, d_2, d_1$. We used the convention that the left subnode has a positive *detail*, and the right subnode has a negative *detail*. (Data precision here to 4 decimal places.)

	s_7	d_7	d_6	d_5	d_4	d_3	d_2	d_1
Sepal.L	0.3672	−0.0547	0.0781	0.0625	0.25	−0.3438	0.25	−0.3625
Sepal.W	0.4236	0.0486	0.0694	−0.0833	0.1111	−0.1944	0.1667	−0.5
Petal.L	0.3594	−0.1719	−0.0313	−0.0625	0.0	−0.0625	0.125	−0.375
Petal.W	0.125	0	−0.125	−0.125	−0.25	-0.25	0	−0.5

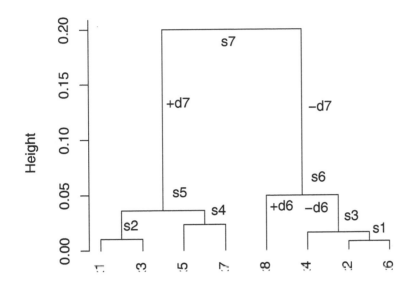

FIGURE 7.2: As Figure 7.1, with *smooths*, s_7, s_6, etc. shown, and with some of the *detail* vectors, d_7, d_6. Details (shown) $+d_7$ and $-d_7$ are associated with offspring branches of node s_7. Details (shown) $+d_6$ and $-d_6$ are associated with offspring branches of node s_6. Details (shown) $+d_5$ and $-d_5$ are associated with offspring branches of node s_5. The situation is analogous to this (although not shown) for nodes s_2, s_5, s_4, s_3, s_1.

reconstruction of the input data. The hierarchy provides a "key" for us to recreate the input data. The total number of values in the dendrogram wavelet transformed data is precisely the same as the number of values in the input data.

7.3 An Object as a Chain of Successively Finer Approximations

From the wavelet-transformed hierarchy we can read off that, say, $x_1 = d_2 + d_5 + d_7 + s_7$ (cf. Figure 7.2). Or $x_8 = d_6 - d_7 + s_7$. These relationships use the appropriate vectors shown (as column vectors) in Table 7.3. Such relationships furnish the definitions used by the inverse wavelet transform, that is, the recreation of the input data from the transformed data.

Thus, the Haar dendrogram wavelet transform gives us an additive decomposition of a given observation (say, x_1) in terms of a degrading approximation, with a variable number of terms in the decomposition. The objects, or observations, are those things which we are analysing and on which we have (i) induced a hierarchical clustering, and (ii) further processed the hierarchical clustering in such a way that we can derive the Haar decomposition. In this section we will look at how this allows us to consider each object as a limit point. The interest or focus lies in our object set, characterized by a set of data, as a set of limit or fixed points.

Using notation from domain theory (e.g. [71]) we write

$$s_7 \sqsubseteq s_7 + d_7 \sqsubseteq s_7 + d_7 + d_5 \sqsubseteq s_7 + d_7 + d_5 + d_2. \tag{7.1}$$

The relation $a \sqsubseteq b$ is read "a is an approximation to b", or "b gives more information than a" (Edalat [72] discusses examples). Adding another term to the very last, or rightmost, term in relation (7.1) gives

$$s_7 \sqsubseteq s_7 + d_7 \sqsubseteq s_7 + d_7 + d_5 \sqsubseteq s_7 + d_7 + d_5 + d_2 \sqsubseteq x_1. \tag{7.2}$$

Every one of our observation vectors (here, for example, x_1) can be increasingly well approximated by a *chain* of the sort shown in relation (7.1) or (7.2), starting with a least element (s_7; more generally, for n observation vectors, s_{n-1}). The observation vector itself (e.g. x_1) is a least upper bound (lub) or supremum (sup), denoted $\sqcup$ in domain theory, of this chain. Since every observation vector has an associated chain, every chain has a lub. The elements of the "rolled down" tree, s_7, $s_7 + d_7$ and $s_7 - d_7$, $s_7 + d_7 + d_5$ and $s_7 + d_7 - d_5$, and so on, are clearly representable as a binary rooted tree, and the elements themselves comprise a partially ordered set (or poset). A *complete partial order* or *cpo* or *domain* is a poset with least element, and such that every chain has a lub. Complete partial orders generalize complete lattices: see [56] for lattices, domains, and their use in fixpoint applications.

7.3.1 Approximation Chain using a Hierarchy

An alternative, although closely related, structure with which domains are endowed is that of spherically complete ultrametric spaces. The motivation comes from logic programming, where non-monotonicity may well be relevant (this arises, for example, with the negation operator). Trees can easily represent positive and negative assertions. The general notion of convergence, now, is related to *spherical completeness* ([212, 96]; see also [115, Theorem 4.1]). If we have any set of embedded clusters, or any chain, q_k, then the condition that such a chain be non-empty, $\bigcap_k q_k \neq \emptyset$, means that this ultrametric space is non-empty. This gives us both a concept of completeness and a fixed point which is associated with the "best approximation" of the chain.

Consider our space of observations, $X = \{x_i | i \in I\}$. The hierarchy, H, or binary rooted tree, defines an ultrametric space. For each observation x_i, by considering the chain from root cluster to the observation, we see that H is a spherically complete ultrametric space.

7.3.2 Dendrogram Wavelet Transform of Spherically Complete Space

Consider analysis of the set of observations, $\{x_i \in X \subset \mathbb{R}^m\}$. Through use of any hierarchical clustering (subject to being binary, a sufficient condition for which is that a pairwise agglomerative algorithm was used to construct the hierarchy), followed by the Haar wavelet transform of the dendrogram, we have an approximation chain for each $x_i \in X$. This approximation chain is defined in terms of embedded sets. Let $n = \text{card } X$, the cardinality of the set X. The Haar dendrogram wavelet transform allows us to associate the set $\{\nu_j | 1 \leq j \leq n-1\} \subset \mathbb{R}^m$ with the chains, as denoted above in equations (7.1) and (7.2).

We have two associated vantage points on the generation of observation i, for all i: the set of embedded sets in the approximation chain starting always with the entire observation set indexed by the set I, and ending with the singleton observation; or the global smooth in the Haar transform, which we will call ν_{n-1}, running through all details ν_j on the path, such that an additive combination of path members increasingly approximates the vector x_i that corresponds to observation i. The two associated views are, respectively, a set of sets; and a set of vectors in $\mathbb{R}^m$. We recall that m is the dimensionality of the embedding space of the observations. The two associated views of the (re)generation of an observation both rest on the hierarchical or tree structuring of our data.

7.4 Generating Faces: Case Study Using a Simplified Model

7.4.1 A Simplified Model of Face Generation

Consider a very simplified model of face recognition, providing a "toy problem" from which some important conclusions will be drawn. Representation or encoding "takes the strain" of this approach, so we need to have that addressed as a matter of priority. For the link with human neural encoding of faces, [232] is a useful starting point. A "perceptual face space" is at issue in [232] and this author proceeds to point to limits of Euclidean embedding of perceptual face spaces, and instead proposes arguments in favour of ultrametric embedding. Therefore [232] is a very useful prolegomenon for all of this work.

We codify our simplified and stylized faces in an analogous way to the encoding often used in the processing of real faces [250]. We use [245] and associated software in R, and also the results presented here that are based on an implementation of Chernoff [48] in S-Plus. We will scale data such that all attributes are in the interval $[0, 1]$. We use 15 attributes for a face, as follows: 1, area of face; 2, shape of face; 3, length of nose; 4, location of mouth; 5, curve of smile; 6, width of mouth; 7, 8, 9, 10, 11, location, separation, angle, shape and width of eyes; 12, location of pupil; 13, 14, 15, location, angle and width of eyebrow.

Here we define our faces as randomly generated, uniformly on the 15 required attributes. Our nine faces in 15-dimensional space are first preprocessed to have all attributes with values on the interval $[-1, +1]$. Following this normalization, a Ward minimum-variance hierarchical clustering of the faces is carried out (see Figure 7.3). We may depict our cases, here faces, in an appropriate way; Figure 7.4 just shows this.

Then a Haar dendrogram wavelet transform was applied. This provides us with a hierarchical decomposition of our given faces, such that we can reconstruct any face from these components. The components, as described in Section 3.6.3, take the overall smooth and successively add or subtract the detail data at a resolution level after the next, as the dendrogram tree is descended from the root node, corresponding to the overall smooth, to the singleton node, corresponding to our selected face. The result of the wavelet transform is shown in Figure 7.4, where detail coefficients and the smooth are depicted as faces.

By proper combination of smooth and details (in Figure 7.4), each one of the faces in Figure 7.3 can be exactly reconstructed. Note that what we have here are mappings of data sets onto the facial representations, which means that the data that we calculate with are encodings of these facial representations. We have a well-defined and unique procedure for (i) decomposing or "peeling away" the input data to yield the transformed data; and (ii) a recomposition, allowing regeneration of the input data. So, for example, we have the following:

- face 3 = smooth − detail at level 8 + detail at level 7 + detail at level 1.

- face 8 = smooth + detail at level 8 − detail at level 5 − detail at level 3 − detail at level 2.

The smooth in Figure 7.5 is the sum of all faces. (Interestingly, the city of Sydney has determined "real life" average faces, involving a great number of people. These average faces are identical to sums, modulo scaling. See [233].)

When the dendrogram is "balanced" or "symmetric" [166], the smooth is, to within a

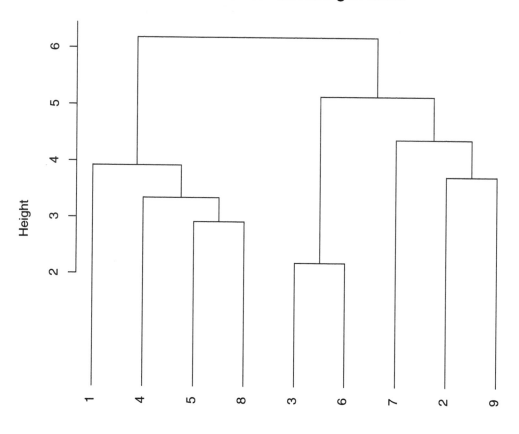

FIGURE 7.3: Hierarchical clustering of our nine faces, with the non-singleton nodes labelled N1, N2, etc.

Our input set of 9 faces

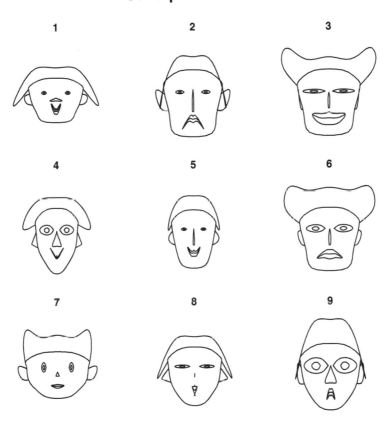

FIGURE 7.4: A Chernoff face depiction of the faces.

**1: Smooth (at root node 8),
2:9: Details (at non–singleton nodes 7 to 1)**

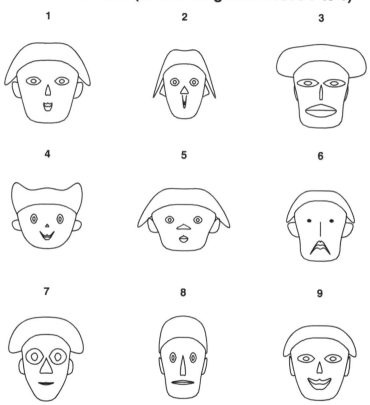

FIGURE 7.5: With reference to Figure 7.3, the overall smooth and the succession of details are displayed. Derived from the wavelet transform of the dendrogram.

constant, the (unweighted) mean object; and the path traversed in the dendrogram, our "key" to reconstituting a face, has approximately $\log n$ steps on it.

7.4.2   Discussion of Psychological and Other Consequences

As noted, the overall smooth, and start point for the reconstruction of any object from the hierarchically represented information, can be to within a constant the mean object. This approach uses a hierarchy as a "key" to the generative mechanism for an object. This approach is therefore a norm-referenced one.

In [89], it is found that norm-referenced encoding of human faces is a more likely mechanism in facial recognition, compared to example-based encoding. The former is with reference to an average or norm, whereas the latter is relative to prototypical faces. This is reinforced in [136]: "The main finding was a striking tendency for neurons to show tuning that appeared centered about the average face". They suggest that norm-referencing is helpful for making face recognition robust relative to viewing angle, facial expression, age, and other variable characteristics. Finally, it is suggested [136]: "Norm-based mechanisms, having adapted to our precise needs in face recognition, may also help explain why our face recognition is so immediate and effortless".

A wide range of experimental psychology results are presented by [60] to support the link between norm-referenced reasoning and unconscious reasoning, on the one hand, contrasted with the link between prototype-referenced reasoning and conscious thinking, on the other hand. We will pursue some discussion of these links since they provide a most consistent backdrop to this methodology.

Encoding of information is fundamental. "Thinking about an object implies that the representation of that object in memory changes." Furthermore, "information acquisition" remains crucial for either form of thought, conscious or unconscious. Further description of how information theory perspectives and accompanying methodology fit fully with mental processes, and especially the role of our unconscious mental processes, is in Annex 1 of Chapter 8.

Dijksterhuis and Nordgren [60] point to how conscious thought can process between 10 and 60 bits per second. In reading, one processes about 45 bits per second, which corresponds to the time it takes to read a fairly short sentence. However the visual system alone processes about 10 million bits per second. It is concluded from this that the conscious thinking process in humans is very low, compared to the processing capacity of the entire human perception system.

7.5   Complexity of an Object: Hierarchical Information

In the context of the face case study, we have considered the following, where m is the number of unique face attributes, and L is the maximum object (i.e. face size or total number (non-unique) of attributes).

In the face case study, the representation was m-length and real. Let each real value be discretized into P intervals. Each face then is described by a boolean Pm-length representation. Directly generating one face requires 2^{Pm} operations. A random face, assuming uniformity, has probability 2^{-Pm}. A Shannon information measure of the object is Pm bits.

Now we change the context, and assume that we have a hierarchical structuring of the set of n objects considered. Directly generating one object requires $O(n)$ operations, and worst case $n - 1$. Each operation is of linear computational complexity in the representation used. A random object, assuming uniformity, has probability n^{-1}. A Shannon information measure of the object is $\log n$ bits.

Our interest lies in cases where $n \; ; m, L, P$, i.e. the total number of objects considered is very much less than the length of their description.

In practice, given a natural macroscopic object class, n may be small, whereas we can go to great lengths to characterize the objects in terms of precision or description length. So the computability of the object is likely to be far more tractable, given this approach based on the hierarchical coding of information.

7.6 Consequences Arising from This Chapter

This approach has been inspired by algorithmic information (or Kolmogorov complexity) that considers a single, finite object and, more particularly, the length of the shortest binary program from which the object can be effectively reconstructed. As a tool, we used a novel wavelet transform on a hierarchy to provide a layer-by-layer reconstruction of the object, starting from an average object (under certain circumstances, a mean object).

Significant challenges face us in regard to how we understand and process objects, as noted by Brooks [39]. A solution that we propose from the work described in this chapter is to explore further "hierarchical coding systems" (this characterization is used in [115]) of the sort used in this work. We have described in this chapter how this can be done, using a practical, simplified case study.

We have shown, theoretically and in the case study presented, that we can generate faces from faces, with a global sum (or average) face as our starting point.

The generation procedure is of average complexity proportional to $\log n$, and worst case $O(n)$, when we are dealing with n objects (here, faces).

This work is consistent with Hawkins [94], who in a machine learning perspective, concludes that: "We have recognized a fundamental concept of how the neocortex uses hierarchy and time to create a model of the world and to perceive novel patterns as part of that model."

Anderson [6] remarks on how "it may be the case that the unique reach and power of human ... intelligence is a result not so much of a unique ability to perform complex, symbolic cognition in abstraction from the environment, but is rather due in large measure to the remarkable richness of the environment in which we do our thinking." He elaborates on this as follows. A central role is played "by persisting institutions and practices in supporting the possibility of high-level cognition. In cognitive science such structures are called scaffolds; a scaffold, in this sense, occurs when an epistemic action results in some more permanent cognitive aid – symbolic, or social-institutional." So "we do very complex things, e.g., building a jumbo jet or running a country 'only indirectly – by creating larger external structures, both physical and social, which can then prompt and coordinate a long sequence of individually tractable episodes of problem solving, preserving and transmitting partial solutions along the way' [49]. These structures include language, especially written language, and indeed all physically instantiated representations or cognitive aids ... Such scaffolds allow us to break down a complex problem into a series of small, easy ones, ... Not just symbol systems, but social structures and procedures can sometimes fill a similar role."

All of this is exciting, but it rests on a fundamental bedrock of representation in the sense of data encoding, together with composition operators defined on these codes. We require, as a *sine qua non* for this work, a data encoding scheme (i) preferably of small, finite length, (ii) capable of being efficiently (low-order polynomial) converted into a display, and (iii) capable of being efficiently (low-order polynomial) determined from a real-world exemplar of the object.

Mainstream physics proceeds by analysing the ever smaller and ever larger. Mainstream computer science has its point of departure in the necessary finiteness of that which is computed. The feasibility of this computer science perspective is based on our finiteness as human beings. An interesting example from [61], discussed in [198], is to consider a person monitored by a video camera for their entire life. The amount of data, for 70 years or 2.2×10^9 seconds, is to an approximation 27.5 terabytes. Let us pose the question of the complexity of a human life, expressed as this particular 27.5 terabytes of information. In a similar vein, the work of Shakespeare, according to [42], amounts to under 1 million words, and can be spoken in 70 hours.

A further supporting view, for music and literary works, is as follows. Basing himself approvingly on a publication by R. Kolisch in 1943, the musical and cultural theorist Theodor Adorno [2] considered "the basic characters to which the types of Beethoven's tempi correspond. In this way, [we arrive] at a discrete number of such basic characters and tempi. At first, the result is shocking; it seems a bit mechanistic and overly mathematical in relation to Beethoven's gigantic oeuvre. But if you turn the tables, ... you will find that great ... music actually bears some resemblance to a puzzle. The movements of the greatest composers are based on a discrete number of *topoi*, of more or less rigid elements, out of which they are constructed. ... Music represents itself as if one thing were developing out of the other, but without any such development literally occurring. The mechanical aspect is covered up by the art of composition". Adorno's discussion continues with a reference to a similar picture in relation to how "Similarly, with a certain amount of naïveté, the great philosophical systems beginning with Plato have had recourse again and again to such mechanical means".

The perspective, based on some hierarchically structured, appropriate representation or encoding of our object family, and an associated algebra, is that it is so much easier to grow the object! Algorithmic complexity traditionally is related to the length (or size) of the object. For us, algorithmic complexity is related to the size of the object class, rather than to the size of the object. Such a perspective is not a replacement for the algorithmic information view. It is simply a different view.

In physics, the pursuit of the ever smaller and ever larger, notwithstanding finite and discrete limits, makes the computability of physical objects difficult and problematic. On the other hand, the finitary computer science view presented in this work, based on hierarchical coding, is eminently tractable and allows natural and artefact objects to be computable.

8

Geometry and Topology of Matte Blanco's Bi-Logic in Psychoanalytics

8.1 Approaching Data and the Object of Study, Mental Processes

In this chapter, the key theme is that of mathematical representations of Matte Blanco's bi-logic, based on metric space and ultrametric or hierarchical topology. This combines both methodological insight and perspectives on practical application.

The Chilean psychoanalyst Ignacio Matte Blanco developed theories of subconscious processes and their relationship with conscious reasoning. This work, especially the book *The Unconscious and Infinite Sets: An Essay in Bi-Logic* published in 1975 [33], uses mathematical perspectives. The close alignment between psychoanalysis and mathematical reasoning is at issue in this chapter. In the preface to his book, Matte Blanco said: "I am firmly convinced that a reformulation of psycho-analysis in terms of logico-mathematical concepts will profoundly revolutionise philosophical thinking and greatly change some of the basic tenets upon which Western philosophy and science rest, and that this book is an attempt, however imperfect, at such a reformulation" [33, p. xxv]. For more discussion on Matte Blanco's work in psychoanalysis, including some indications of mathematical viewpoints, see [13].

Before we approach the data-based study of the human mind, it will be very useful to note how important and how feasible it is to integrate analytic, synthetic) mental processes with the objects of these, our human mental processes. This is surely all the more essential when the objects of mental processes happen to be those very same mental processes.

Jean-Paul Benzécri's approach to data analysis, pursued in Paris (Université Pierre et Marie Curie, Paris 6) between the 1970s and 1990s was often referred to as the French school of data analysis. It displayed both cohesiveness of theoretical underpinning and breadth of applications. Benzécri's early data analysis motivation, in Rennes to begin with and subsequently in Paris, came from linguistics.

The data analysis "platform" [177], with its comprehensive reasoning infrastructure, is based on the correspondence factor analysis methodology (with which is very closely associated the related method of agglomerative hierarchical clustering), and other multivariate data analysis, and more broadly, statistical analysis methods. Examples of these will be seen below.

Correspondence analysis was initially proposed as an inductive method for analysing linguistic data. From a philosophy standpoint, correspondence analysis simultaneously processes large sets of facts, and contrasts them in order to discover global order; and therefore it has more to do with synthesis (etymologically, to synthesize means to put together) and induction. On the other hand, analysis and deduction (viz., to distinguish the elements of a whole; and to consider the properties of the possible combinations of these elements) have become the watchwords of data interpretation. It has become traditional now to speak of data analysis and correspondence analysis, and not "data synthesis" or "information synthesis". In every practical way, these are most essential, data integration and information synthesis.

8.1.1 Historical Role of Psychometrics and Mathematical Psychology

The related areas of psychometrics and mathematical psychology in data analytics are insightful, important and most relevant for such areas as inductive reasoning and information fusion. A short historical overview is given here. Benzécri's short historical outline of data analytics [26] offers insights into multivariate data analysis and multidimensional statistics.

Psychometrics made multidimensional or multivariate data analysis what it has now become, namely, "search by induction of the hidden dimensions that are defined by combinations of primary measures". Psychometrics is a response to the problem of exploring areas where immediate physical measurement is not possible (e.g. intelligence, memory, imagination, patience). Hence a statistical construction is used in such cases – "even if numbers can never quantify the soul!" [26].

While it is now part of the history of data analysis and statistics that around the start of the twentieth century interest came about in human intelligence, and an underlying measure of intelligence, the intelligence quotient (IQ), there is a further link drawn by [26] in tracing an astronomical origin to psychometrics. Psychophysics, as also many other analysis frameworks such as the method of least squares, was developed in no small way by astronomers: the desire to penetrate the skies led too to study of the scope and limits of human perception, and hence psychometrics.

8.1.2 Summary of Chapter Content

This chapter is set out as follows. First, it will be shown how versatile and flexible, data-driven narrative analysis in text can be carried out. The examples involve both partially structured film script text, and direct use of unstructured text in a novel.

In Section 8.2 salient and important points will be taken from Matte Blanco's work. These points are of direct relevance in what follows.

Metric properties are discussed in Section 8.3. Consider, for example, the natural (from a human visual standpoint) Euclidean distance. The Euclidean distance between Paris and London, for example, can be calculated from their planar projection coordinates, with curvature of the Earth, or the presence of the Channel, etc., ignored.

Section 8.4 then deals with the ultrametric, a special metric. A tree distance is an ultrametric – a distance that is defined on a rooted tree. Thus what we need to know for the ultrametric distance between two points on two different branches is how to get back to the common stem of these branches.

To conclude this chapter, there are two additional annexes describing very interesting follow-on issues. Firstly, there is discussion of how human thought processes might be expressed in terms of computation, both for conscious and for unconscious mental processes. Secondly, there is some further discussion of practical aspects of text analysis that is motivated by, and makes use of, the insights brought about by Matte Blanco's work.

8.1.3 Determining Depth of Emotion, and Tracking Emotion

Continuing from Chapter 1, in particular, a brief discussion is given of the mapping of emotion. This involves depth of emotion, and tracking of emotion, using the movie *Casablanca* and Gustave Flaubert's nineteenth-century novel, *Madame Bovary*. See the book's website for the relevant data, and the R software used.

Following [187], we look at emotional interaction in *Casablanca*, using dialogue (and dialogue only) between main characters Ilsa and Rick, having selected this dialogue from the 11 scenes involving both of these protagonists (scenes 22, 26, 28, 30, 31, 43, 58, 59, 70, 75 and 77). Then we look at all of the text in Part 2, Chapters 9–12 of *Madame Bovary*. This concerns the three-way relationship between Emma Bovary, her husband Charles, and her lover Rodolphe Boulanger.

The same methodology is used as described in Chapter 1: cross-tabulation of word sets, comprising the entire universe of discourse, and including function and grammar words that characterize textual "texture". The latter, function words, can be found to be very useful for expression of emotion. On the latter issue, using function or grammar, or form or tool, words, see [160, 203].)

The Rick–Ilsa dialogue starts and ends as follows:

```
1, RICK, "-- Hello, Ilsa."
2, ILSA, "Hello, Rick."
3, ILSA, "-- This is Mr. Laszlo."
4, RICK, "How do you do?"
5, ILSA, "I wasn't sure you were the same. Let's see, the last time we met --"
6, RICK, "-- It was La Belle Aurore."
7, ILSA, "How nice. You remembered. But of course, that was the day the
Germans marched into Paris."
8, RICK, "Not an easy day to forget."
9, ILSA, "No."
10, RICK, "I remember every detail. The Germans wore gray, you wore blue."
11, ILSA, "Yes. I put that dress away. When the Germans march out, I'll wear
it again."
...
150, RICK, "You better hurry, or you'll miss that plane."
```

The tracking of emotion is carried out through representative words "darling" and "love" in Figure 8.1. The importance of these words is signalled very clearly in the overall view of the evolution of emotion that is displayed in Figure 8.2.

Attention is now turned to the three-way (Emma, Charles, Rodolphe) relationships in *Madame Bovary*. In order to be applied in a very general, robust, manner to textual description such as this, successive text segments of 20 lines each are used.

Figure 8.3 presents an interesting perspective that can be considered relative to the original text. Rodolphe is emotionally scoring over Charles in text segment 1, then again in segments 3–6. In text segment 7, Emma is accosted by Captain Binet, giving her qualms of conscience. Charles regains emotional ground with Emma through Emma's father's letter in text segment 10, and Emma's attachment to her daughter, Berthe. Initially the surgery on Hippolyte in text segment 11 draws Emma close to Charles. By text segment 14 Emma is walking out on Charles following the botched surgery. Emma has total disdain for Charles in text segment 15. In text segment 16 Emma is buying gifts for Rodolphe in spite of potentially making Charles indebted. In text segments 17 and 18, Charles's mother appears, with whom Emma has a difficult relationship. Plans for running away ensue, with pangs of conscience for Emma, and in the final text segment Rodolphe is talking himself into refusing to leave with Emma.

Figure 8.4 displays the evolution of sentiment, expressed by (or proxied by) the terms "kiss", "tenderness", and "happiness". We see that some text segments are more expressive of emotion than others.

In these case studies, which track the emotion between Ilsa and Rick in *Casablanca*, and between Emma, Rodolphe and Charles in *Madame Bovary*, subplots in both narratives are also determined. Further discussion of these case studies can be found in [187].

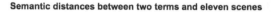

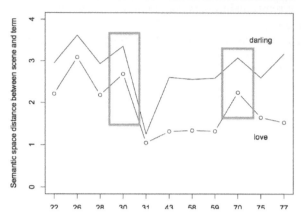

FIGURE 8.1: *Casablanca.* In the full-dimensionality factor space, based on all interrelationships of scenes and words, the distance was determined between the word "darling" in this space, with each of the 11 scenes in this space. The same was done for the word "love". The semantic locations of these two words, relative to the semantic locations of scenes 30 and 70, are highlighted with boxes. In these two scenes, there is a pronounced ebbing of emotion.

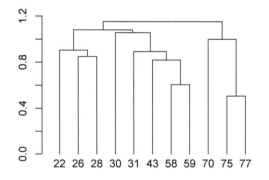

FIGURE 8.2: *Casablanca.* Hierarchical clustering, sequence or chronologically constrained, of the 11 scenes with dialogue, and only dialogue, between Ilsa and Rick. See how the big changes in scenes 30 and 70 are indicated in Figure 8.1.

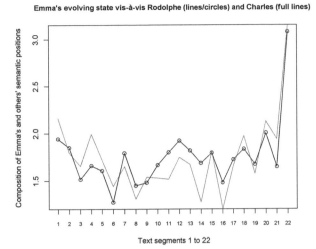

FIGURE 8.3: *Madame Bovary.* The relationship between Emma and Rodolphe (lines/circles, black) and between Emma and Charles (full line, grey) is mapped out. The text segments encapsulate narrative chronology, which maps approximately into a time axis. Low or small values can be viewed as emotional attachment.

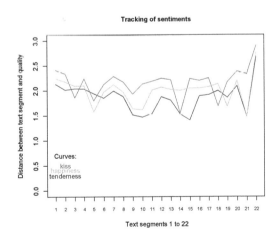

FIGURE 8.4: *Madame Bovary* novel. The ebb and flow of emotion is displayed. A low value is a strengthening of emotion, and a high value is a weakening of emotion. A low value of the emotion, expressed by the words "kiss", "happiness" and "tenderness", implies small distance to the text segment. The chronology of sentiment tracks the closeness of these different sentimental terms relative to the narrative, represented by the text segment. Terms and text segments are vectors in the semantic, factorial space, and the full dimensionality of this space is used.

8.2 Matte Blanco's Psychoanalysis: A Selective Review

Matte Blanco's work is concerned with conscious reasoning and subconscious thought processes. His great achievement is to develop a cognitive model that embraces both.

We start with a selective review of Matte Blanco's work in order to set out the key terms and key thrusts of his work, such that there will be a strong resonance with the mathematical concepts and mathematical or computational processing at issue in this chapter.

Matte Blanco's *The Unconscious as Infinite Sets* (originally published in 1975; see [33]) was, according to the author, "written for psycho-analysts as well as for mathematical philosophers" and is described in Eric Rayner's foreword as "undoubtedly [his] most fundamental work".

We begin by summarizing particularly salient aspects of Matte Blanco's work under the points laid out as follows. That will help in showing how the various points can be seen in mathematical terms. Quotations in the following are from [33].

1. Matte Blanco related his work to Freud's conscious or unconscious.

   - Relative to Freud's work, Matte Blanco had it "largely reformulated in terms of symmetry and asymmetry".

   - For him, these were "two kinds or modes of being rather than of existence". The interplay of symmetry and asymmetry is the focus of Matte Blanco's work.

   - The upshot of this was that Matte Blanco arrived at what he termed a "bi-logical" system or "bi-logic".

   There are "two fundamental types of being which exist within the unity of every man: that of the 'structural' id (or unrepressed unconscious or system unconscious or symmetrical being) which becomes understandable with the help of the principle of symmetry; and that visible in conscious thinking, which can roughly be comprehended in Aristotelian logic."

   Freudian consciousness and unconsciousness are reformulated in terms of symmetrical and asymmetrical modes of being. Note that this is not a Freudian "rational–irrational" polarity but rather, on the side of the symmetric mode of being, the "unrepressed unconscious", or what is "the unconscious by its own nature or structural unconscious". As seen in the development of the theory of Matte Blanco, "[i]t is an attempt at putting in logico-mathematical terms the findings of Freud".

2. Symmetrization is a principle which, as shown by Matte Blanco, helps in understanding:

   - schizophrenia, and clinical treatment and practice
   - metaphor and other figures of speech
   - jokes, and disjunctions or abrupt change in discourse
   - emotion and emotionally loaded language
   - dream
   - poetry, literature, art
   - subconscious

- •the "structured" id (or unrepressed unconscious)
- •our system unconscious (or symmetrical *being*).

What is especially considered by him is "pure symmetry" at the deepest level, that is, at "the level of *being*, in contrast to the level of *happening*" (emphasis in original).

For Matte Blanco, "[s]ymmetrical being is the normal state of man." From it, its counterpart (in a way, its dual), "consciousness or asymmetrical being emerges" and "makes attempts at describing it" (i.e. the primary experience of symmetric being). We can justifiably consider "symmetrical logic" as the framework of this description by asymmetrical thought that, in some manner, is derived from primary symmetrical thought. He says: "the most central trait" of symmetrical being "is the peculiar (extensive) use of symmetrical relations" – hence, "the symmetrical mode of being or symmetrical mode."

3. Within a class of things as conceptualized by the thinking person, there is perfect equivalence of class members, implying the following:

    - •no contradiction
    - •absence of negation
    - •displacement
    - •space and time vanish
    - •no relations of contiguity
    - •arising from the latter, no order.

4. How a class is defined in practice, or is known to the thinking person, is described in these terms.

    - •Because, as elaborated on by Matte Blanco, one class member is – in terms of class membership – indistinguishable from another class member, "the unconscious does not know individuals but only classes or propositional functions which define the class".
      Further, "The only unity for the (symmetrical) unconscious is the class or set, in which all individuals belonging to it are included. The unconsciousness cannot, therefore, deal with parts, except by treating them as classes or sets."
    - •"Consciousness ... when confronted by a whole class can only consider it in two ways: either it focuses on the limits (or definition) of the class, that is, on those precise features which characterize it and distinguish it from all other classes, or it concentrates on the individuals which form the class."

5. A class comes about through condensation: "two impulses which appear incompatible in Aristotelian logic and their union in one expression, ... is accomplished in condensation".

6. The principle of generalization relates different classes.

    - •We assume various classes.
    - •Then "the principle of generalization and the principle of symmetry" are both taken for their explanatory capability in regard to classes.
    - •In this way, the "generalizing part [in the human] leads to symbols", since symbols arise out of knowledge of, or awareness of, classes.

•Classes are structured as what might be called "bags of symmetry" (in quotation marks in the original), and also "levels".

7. Counterposed to the symmetrical principle in Matte Blanco is the asymmetrical principle.

   •It is visible in conscious thinking.

   •It can roughly be comprehended in, or expressed through, Aristotelian logic.

   •"Asymmetrical being ... perceives reality as divisible or formed by parts and, as such, related to spatio-temporality".

   •Symmetrical being can by known only through the glass or prism of asymmetrical being: "Thinking requires asymmetrical relations. So does consciousness."

8. Quantifying the symmetrical.

   •"Symmetrical being alone is not observable in man." Even delineating it is "already an asymmetrical ... activity".

   •In regard to emotion, the "magnitude of emotion" is understood in terms of "the proportion between symmetrical and asymmetrical thinking".

   •"[U]nconscious psychological events are not intrinsically immeasurable" although compared to a physical event being susceptible to just one measurement, instead with unconscious events it is a matter of being susceptible to infinite measurement. Infinite measurement may be understood on the basis of the Cantor argument whereby a whole set, when in a bijection with a part of this same set, implies the same countable infinite cardinal for both whole and part sets.

   •"By making the individual identical to the class, the principle of symmetry, *as seen from an asymmetrical point of view*, leads to the infinite set".

   •"We must ... keep in mind the possibility that if things are viewed in terms of multidimensional space, symmetrical being can actually unfold into an infinite number of asymmetrical relations."

9. In free recall, and in other areas besides such as in literature, words are tracers for expressing what lies behind.

   •"Consciousness cannot exist without asymmetrical relations, because the essence of consciousness is to distinguish and to differentiate and that cannot be done with symmetrical relations alone."

   •"Symmetrical being is translated into asymmetrical terms by means of words. *Words (i.e. their meanings) are the asymmetrical tools of the translating-unfolding function*" (emphasis in the original).

   •We have that "words, abstract things, fulfill the function of differentiating between concepts and also between other things. They are bound to be, therefore, highly asymmetrical in their structure."

   •To the foregoing can be added that Text is the "sensory surface" [153], formulated in statistical and computational terms of the underlying semantics in Section 2.5 above. In Annex 2 of this chapter, there will be further motivation as to why words are a good starting point for further analysis and how this can even go towards accessing aspects of underlying symmetrical being.

Thus far, various central themes have been selected from Matte Blanco. This leads us to a conclusion drawn by Lauro-Grotto [129] that directly follows from Matte Blanco: "here comes my observation: the structural unconscious, in the way it is reformulated by Matte Blanco, the symmetric mode – all this is homologous to an ultrametric structure. The generalization principle reflects the hierarchical arrangement in which all the stimuli (or concepts) are perceived as belonging to classes, and the classes are clustered into super-classes of increasing generality. Finally, a single omni-comprehensive class is generated."

Lauro-Grotto [129] points to how equi-similar (or equi-distant) stimuli or concepts indicate an ultrametric (or hierarchy, or tree) topology.

It will be both observed and demonstrated in Sections 8.3 and 8.4 how the laying out by Matte Blanco of the symmetric and asymmetric principles leads in a very natural way to an ultrametric topology as a representational model.

8.3 Real World, Metric Space: Context for Asymmetric Mental Processes

Our real ambient space, and an ordered timeline, are fundamental to our human, conscious mental processes.

A metric perspective is very appropriate for the usual three-dimensional world around us, to which we can add time as a fourth dimension. The real number line is then appropriate for any (space or time) dimension of this world. For geometric modelling, Cartesian coordinates can be availed of, where this appellation comes from mathematician and philosopher, René Descartes (1596–1650). While other metrics can be important (e.g. in curved space-time worlds), correspondence analysis uses, through and through, the Euclidean distance. Very informally expressed, this is the distance "as the crow flies".

As discussed in previous sections, such geometry is good and fitting for semantic analysis. Text segments are defined as the average of their constituent words, and words are defined as the average of their constituent text segments. Thus, semantic analysis is based on a metric embedding. A high point of such semantic analysis, using metric embedding, can be seen in Bourdieu's use of correspondence analysis. The renowned social scientist, Pierre Bourdieu (1930–2002), used multiple correspondence analysis for mapping survey data into a metric, latent semantic space. This was done in various works including his book, *Distinction: A Social Critique of the Judgement of Taste*, published in French in 1979, and in English in 1984. Bourdieu's work relates to social and aesthetic preferences, cultural and educational capital.

For a metric context, the natural geometric ordering is on the real line, whereas in the ultrametric case the natural ordering is a hierarchical tree.

Khrennikov [119] notes how the real number system is used for measurement in real-world, physical spaces. But ultrametric distances are appropriate, Khrennikov holds, for mental spaces. See [114, 115, 116, 118] for a great amount of work studying dynamical processes in ultrametric spaces. Thus, Khrennikov lays the basis for unconscious information processing. Khrennikov [119] notes the following: "The idea that physical and mental spaces have essentially different geometries was discussed already by Aristotle. He emphasized continuity, infinite divisibility, and connectivity of the physical space. At the same time he presented motivations that the mental space is discrete, hierarchic, and totally disconnected. The latter matches perfectly with the modern notion of a totally disconnected topological space."

In practical and operational settings we can take text, or text transcribed from voice, or survey or questionnaire results, or other measured data on our observables, together with the characteristics of our subjects, or observables, and map such data into a semantic space that is metric (because, by design, the factor space, or latent semantic space, is a high-dimensional metric space). Then we can proceed from there to induce a hierarchy, or tree structure.

8.4   Ultrametric Topology, Background and Relevance in Psychoanalysis

Having surveyed Matte Blanco's view of unconscious thought processes expressed as (Matte Blanco's term) symmetry, and conscious reasoning expressed as (again Matte Blanco's term) asymmetry, in this section a basis will be laid out for mathematically modelling these – symmetry, asymmetry – as respectively ultrametric (i.e. metric on a tree or hierarchy) and metric.

As observed by Lauro-Grotto [129, p. 539], the aspect of anomaly modelling via an ultrametric is nicely consistent with Matte Blanco's symmetrical logic: "we know that something similar can actually be experienced in finite space when we look at a very distant three-dimensional structure and we perceive it as though it were a single point. Symmetrization of relationships can therefore be described as a transition from a metric to an ultrametric conceptual organization."

8.4.1   Ultrametric

As a distance *on* a tree, an ultrametric is the closest common ancestor distance.

In metric terms, in an ultrametric space all triangles are either isosceles with small base, or equilateral (see Figure 8.5; cf. Section 1.2.5). From x's point of view, y and z are indistinguishable. In mathematical terms, an ultrametric topology (which we can quite acceptably term a hierarchical topology or a tree topology) has further properties that are recognizably along the lines of Matte Blanco's symmetric mental processes. In so far as a cluster, corresponding to any node of the hierarchical tree, contains a set of objects, these objects are all identically members of this cluster. There are unusual properties. Any member of the cluster can be taken as its centre. As a mathematical ball or hypersphere, the radius equals its diameter. The cluster or ball is topologically open and closed at the same time and this is termed a *clopen* set: the set is closed because objects on its boundary can be members; and it is open because the cluster extremity is defined by what is not a member relative to the external, complement set.

In an ultrametric space all triangles are either isosceles with small base, or equilateral. We have here very clear symmetries of shape in an ultrametric topology. For further discussion on this, see Chapter 3. These symmetry "patterns" can be used to fingerprint data sets. A further line of research is to use such patterns to detect subjective and perhaps even subconscious equivalences. Such work has parallels with the motifs relating to relationship triangles used by Neuman [194, Chapter 9].

It is clear that an ultrametric topology is very different from our intuitive (Euclidean) notions. The most important point to keep in mind is that in an ultrametric space everything "exists" in a hierarchy expressed by a tree.

To conclude, therefore, it has been sought to show how well an ultrametric space models

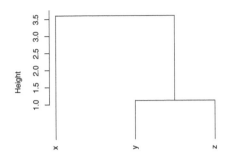

FIGURE 8.5: (Left) The strong triangle inequality defines an ultrametric: every triplet of points satisfies the relationship $d(x, z) \leq \max\{d(x, y), d(y, z)\}$ for distance d. See, by reading off the hierarchy, how this is verified for all x, y, z. In addition, the symmetry and positive definiteness conditions hold for any pair of points. (Right) A depiction of these three points. They form an isosceles triangle with small base. The base is formed by points y, z.

Matte Blanco's symmetry, and how ultrametric space provides a framework for understanding symmetrical being, or a mathematical model of symmetrical being.

8.4.2 Inducing an Ultrametric through Agglomerative Hierarchical Clustering

In the practical and operational context of analysing data, inducing an ultrametric means building a hierarchical clustering from given data. A mapping of metric to ultrametric is achieved by an agglomerative hierarchical clustering algorithm, a well-established approach that depends on a cluster (compactness, connectedness, or other) criterion.

Finally, in this short discussion of hierarchical clustering, to draw yet another link to the work of Matte Blanco, it is noted in Rayner's [208] review of Matte Blanco that the latter's investigation of the *"process of thinking ...* emphasizes the essential centrality of *classificatory* activity at all levels of thought, even in the unconscious".

Figure 8.6 shows a hierarchical clustering. Each non-terminal node in the tree has two child nodes. This is a dendrogram, representing a set of $n - 1$ agglomerations based on n initial data vectors. Non-embedded clusters are, by definition, non-overlapping.

8.4.3 Transitions from Metric to Ultrametric Representation, and Vice Versa, through Data Transformation

Our first basic viewpoint is that a metric modelling and representation is very fitting for real, physical space. The real number system, and the real Cartesian coordinate system, do justice to physical space, and to time. This gives us our four-dimensional physical world of space-time. Newtonian and Einsteinian dynamics (equations of motion, convergence, operation of physical laws in mechanics, fluids, optics, and so on) were developed in the real number system.

Part and parcel of the real number framework is asymmetry, as in the order and indeed total order that can be established by an operator such as "less than or equal to" ($\leq$) on any set of scalar or unidimensional values.

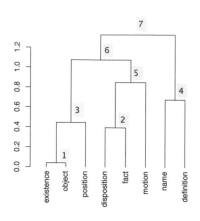 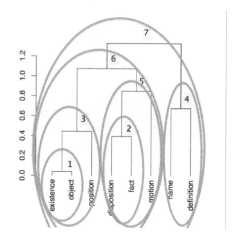

FIGURE 8.6: Hierarchical clustering of eight terms. The data on which this was based are the frequencies of occurrence of the eight nouns in 24 successive, non-overlapping segments of Aristotle's *Categories*. The eight terms comprise the terminal (or leaf) nodes. In the hierarchy these are singleton clusters. The non-terminal nodes (i.e. clusters) are numbered in sequence (the shaded numbers beside the relevant node). If one cuts horizontally through the dendrogram or tree or hierarchy, one gets a partition of the set of eight terms. On the right, embedded clusters are shown with ellipses.

Real, physical space therefore is suitably characterized, represented and modelled by Matte Blanco's concept of asymmetric reasoning. Reciprocally, asymmetric reasoning expresses well the real, physical space of the conscious, self-aware world around us.

With our second basic viewpoint, we come to Matte Blanco's symmetric mental processes, reflecting the human unconscious and/or subconscious. We have alternative number systems, in particular, the p-adic, and their topological equivalent in ultrametric spaces. An ultrametric space is a hierarchical space. Mathematically, the order relation on clusters in a hierarchy is a *partially ordered set*, also termed a *poset*. With reference to the hierarchies of clusters displayed in figures in this chapter, a partially ordered set requires any two clusters to be either one embedded in the other, or, if not, then disjoint (non-overlapping). The embeddedness, or set inclusion, means that we have a natural order on any branch of the hierarchy/tree. But between branches of the hierarchy/tree we have no such natural order.

By transforming metric data to ultrametric, the data are endowed with an ultrametric. The reverse mapping allows various possibilities, including taking a hierarchy into a metric space. In [182] there is discussion of transforming data to become more metric, or to become more ultrametric. There are various contexts to motivate such modifications to the data, for example, to make the data more searchable if information retrieval is the aim, or perhaps to more easily visualize the data in a low-dimensional (two- or three-dimensional) metric space projection.

Such work may be noted in regard to how we have good mathematical apparatus available to understand and suitably express transitions from asymmetric to symmetric reasoning, and vice versa.

8.4.4 Practical Applications

The early chapters of [13] present triangulation, based on Matte Blanco's work, from a mainstream psychoanalysis perspective. This is in relation to many sorts of violent crime.

From a data analysis point of view, in [181] dream reports are mapped through correspondence analysis of semantic content, and then triangular, hence ultrametric, relationships between named individuals in these dream reports are determined. It can be concluded that certain particular patterns emerge from these personal, past and present, relationships, as expressed in these dream report texts. See Section 9.5.1 for further description of this case study.

8.5   Conclusion: Analytics of Human Mental Processes

In a sense, this chapter has been oriented towards a new data analytics, inspired by Matte Blanco, that addresses human mental processes.

Mathematically modelling Ignacio Matte Blanco's principles of *symmetric* and *asymmetric being* in this chapter has objectives such as the following: visualization of concepts developed by Matte Blanco, and hence use as an aid for understanding Matte Blanco's work and imparting it to others; and to provide pointers to where and how such mathematical modelling can be instrumentalized. The ambition is even greater, namely, through Matte Blanco's work to set out and make extensive use of the geometry and the topology of mental processes.

It has been shown how the topology of a rooted tree (i.e. an ultrametric topology) can be used as a mathematical model for the structure of the logic that reflects or expresses Matte Blanco's symmetric being, and hence of the reasoning and thought processes involved in unconscious reasoning or in reasoning that is lacking, perhaps entirely, in consciousness or awareness of itself. Such an ultrametric model corresponds to hierarchical clustering that can be induced on empirical data (e.g. text). For Matte Blanco's asymmetric, real-life, and conscious reasoning, a map of empirical data that captures such perceived and defined objects of observation can be formed such that the semantics can be modelled through Euclidean geometry. This furnishes geometrical representation of the latent semantics. Finally, the passage from metric geometry to ultrametric topology has been discussed, and vice versa. Thus Matte Blanco's bi-logic, in its passage from symmetric to asymmetric (in creativity, when emotion comes to lead behaviour, or when trauma dominates behaviour) or from asymmetric to symmetric (in dream, in repression or in traumatization), in passages both ways between symmetric and asymmetric being, our geometric and topological modelling is enlightening and elucidating.

A few very general remarks follow on the mathematics used in this chapter. In geometry there is a focus on spatial properties, with particular attention given to distance. Topology is the study of shape relations. The topology of a tree, with a root and branches, is of interest to us, and for us this encapsulates hierarchy. In topology, particular attention is given to neighbourhood properties. Informally expressed, we are viewing Matte Blanco's asymmetric mental processes in terms of geometry, and we are viewing symmetric mental processes in terms of (tree, hierarchy) topology.

8.6   Annex 1: Far Greater Computational Power of Unconscious Mental Processes

This annex and the next look at how two aspects follow on from the mathematical models of bi-logic. Firstly, there is one possible explanation for the need for bi-logic in evolution, with a fairly basic information processing viewpoint. Secondly, the practical implications of the mathematical modelling are examined, for finding theoretical support in observed, measured or collected, empirical data.

In this first annex, norm-referenced reasoning and unconscious thought processes are described and contrasted with prototype-referenced reasoning and conscious reasoning.

In [178] a generative theory of information is developed. Given that algorithmic complexity views the complexity of an object as the work required to generate it, the work is described that is needed to generate an object in an ultrametric space as ultrametric algorithmic information. The approach uses a hierarchy as a "key" to the generative mechanism for an object. It is a *norm-referenced* approach.

This leads to further support for the hierarchical model, hence an ultrametric topology, as a model for unconscious thought, as will now be discussed.

In Giese and Leopold [89], it is found that norm-referenced encoding of human faces is a more likely mechanism in facial recognition, compared to example-based encoding. The former is with reference to an average or norm, whereas the latter is relative to prototypical faces.

Leopold et al. [136] reinforce this: "The main finding was a striking tendency for neurons to show tuning that appeared centred about the average face". They suggest that norm-referencing is helpful for making face recognition robust relative to viewing angle, facial expression, age, and other variable characteristics. Finally, they suggest: "Norm-based mechanisms, having adapted to our precise needs in face recognition, may also help explain why our [human] face recognition is so immediate and effortless".

A wide range of experimental psychology results are presented by Dijksterhuis and Nordgren [60] to support the link between norm-referenced reasoning and unconscious reasoning, on the one hand, contrasted with the link between prototype-referenced reasoning and conscious thinking, on the other hand. Some discussion of these links will be pursued since they provide a most consistent backdrop to the Matte Blanco focus in this chapter.

Encoding of information is fundamental. "Thinking about an object implies that the representation of that object in memory changes." Furthermore, "information acquisition" remains crucial for either form of thought, conscious or unconscious.

Dijksterhuis and Nordgren [60] point to how conscious thought can process between 10 and 60 bits per second. In reading, one processes about 45 bits per second, which corresponds to the time it takes to read a fairly short sentence. However the visual system alone processes about 10 million bits per second. It is concluded from this that the conscious thinking process in humans is very low, compared to the processing capacity of the entire human perception system.

A hypothesis is advanced here as to why human thinking includes unconscious as well as conscious thought. Namely, we note that *conscious* reasoning is slow compared to the vastly more efficient and dramatically faster processing speed of *unconscious* thought processes. Dijksterhuis and Nordgren [60] also point to how unconscious thought is less precise and carries no order, including chronological, information. These aspects have already been noted in Matte Blanco's symmetry and the ultrametric interpretation.

Conscious thought therefore is both limited and limiting. A small number of foci of interest ("only one or two attributes") have to take priority. There are inherent limits to conscious thought as a result. As a result of limited capacity, "conscious thought is guided by expectancies and schemas": Limited capacity therefore goes hand in hand with use of stereotypes or schemas. "people use ... stereotypes (or schemas in general) under circumstances of constrained processing capacity ... [While] this [gives rise to the conclusion] that limited processing capacity during *encoding* of information leads to more schema use, [current work proposes] that this is also true for thought processes that occur after encoding. ... people stereotype more during impression formation when they think consciously compared to when they think unconsciously. After all, it is consciousness that suffers from limited capacity."

Dijksterhuis and Nordgren [60] go on to say that it may be be considered counterintuitive that stereotypes are applied in the limited capacity, conscious thought, regime. However stereotypes may be "activated automatically (i.e. unconsciously)", but "they are applied *while we consciously think* about a person or group". Conscious thought is therefore more likely to (unknowingly) attempt "to confirm an expectancy already made".

On the other hand, unconscious thought is less biased in this way, and more slowly integrates information. "Unconscious thought leads to a *better organization* in memory", arrived at through "incubation" of ideas and concepts. "The unconscious works ... aschematically, whereas consciousness works ... schematically". And "conscious thought is more like an architect, whereas unconscious thought behaves more like an archaeologist".

Viewed from the perspective of the work discussed in this annex, it can be appreciated that the hierarchical and generative description of an object set is a simple model of unconscious thought. (That it is simple is clear: to begin with, it is static.) The hierarchical and generative description of an object set (cf. [178]) is underpinned by the object set being embedded in an ultrametric topology.

We find that, in this framework, the information content is defined from the size of the object set, and not from any given object. To that extent, therefore, the computational (or generative) potential of unconscious mental processes is far more powerful that that of conscious mental processes, or thinking.

8.7 Annex 2: Text Analysis as a Proxy for Both Facets of Bi-Logic

Both conscious or asymmetric reason, and unconscious or symmetric reason, are facets of bi-logic according to Matte Blanco. What he means is that both play a role at different times, that these roles are often complementary, and that the interplay of the two separate domains can be very revealing and instructive.

This annex section addresses the plausibility of appreciable analysis of content of thought processes based on interrelationships that in turn are frequencies of co-occurrence data. Text will be used as a proxy for underlying thinking, reasoning, conscious phenomena and also, every bit as much, representative of the underlying emotional, dreaming, or other unconscious mental processes. What is being sought is an approach that is deployable and hence usable in practice.

Words are a means or a medium for getting at the substance and energy of a story, notes McKee [153, p. 179]. Ultimately sets of phrases express such underlying issues (the "subtext", as expressed by McKee) as conflict or emotional connotation. Change and evolution are inherent to a plot. Human emotion is based on particular transitions in thought. So this establishes well the possibility that words and phrases are not only taken literally but can appropriately capture and represent such transition. Text, says McKee, is the "sensory surface" of a work of art (counterposing it to the subtext, or underlying emotion or perception).

Simple words can express complex underlying reality. Aristotle, for example, employed words in common usage to express technically loaded concepts [171, p. 169], as did Freud.

Rayner [208] notes the following: "The unconscious largely deals not with particular logically asymmetrically locatable subjects and objects, but with abstract attributes, qualities or notions. Put in another way, these propositional functions are adjectival and adverbial; they lie behind verbal nouns: lovingness, frighteningness and so on." Such words, he notes, are "abstract class attributes, notions or conceptions" and "are the equivalent of the propositional functions of the class".

This has an immediate bearing on the words used in unconscious processes. Rayner [208] notes the "propositional functions or abstract attributes" or "predicate thinking" that underlie the unconscious as fundamental constituents. He also briefly exemplifies this through clinical work in schizophrenia and child abuse by adults.

One could of course deal with other units of thinking, or reasoning, or unconscious processes, other than through words. Chafe [45], in relating and establishing mappings between memory and story, or narrative, considered the following units.

1. *Memory* expressed by a *story* (memory takes the form of an "island"; it is "highly selective"; it is a "disjointed chunk"; but it is not a book, nor a chapter, nor a continuous record, nor a stream).

2. *Episode*, expressed by a *paragraph*.

3. *Thought*, expressed by a *sentence*.

4. A *focus*, expressed by a *phrase* (often these phrases are linguistic "clauses"). Foci are "in a sense, the basic units of memory in that they represent the amount of information to which a person can devote his central attention at any one time".

The "flow of thought and the flow of language" are treated at once, the latter proxying the former, and analysed in their linear and hierarchical structure as described in other essays in the same volume as [45, 44].

In [179, 180] the following question is addressed: Can we attempt to separate out good proxies for symmetrical logic and for asymmetrical logic? To do this, a great number of texts are taken, relating to literature, technical writing, and after-the-fact reporting on unconscious thought processes.

9

Ultrametric Model of Mind: Application to Text Content Analysis

9.1 Introduction

In the previous chapter, it was discussed how Matte Blanco's work linked the unrepressed unconscious (in the human) to symmetric logic and thought processes. It was noted how ultrametric topology provides a most useful representational and computational framework for this. We now look at the extent to which we can find ultrametricity in text. We use coherent and meaningful collections of nearly 1000 texts to show how we can measure inherent ultrametricity. On the basis of our findings we hypothesize that inherent ultrametricty is a basis for further exploring unconscious thought processes.

Any agglomerative hierarchical procedure (cf. [25, 24, 137, 163, 167]) can impose hierarchical structure. Our first aim in this work is to assess the inherent extent of hierarchical or ultrametric structure.

We take a large number of meaningful texts in order to see how they can be distinguished and/or what other conclusions can be drawn, in regard to their inherent ultrametricity or hierarchical structure.

The procedure is as follows.

1. Meaningful component parts of texts are used, such as chapters, reports, tales, or very approximately similar sized segments of contiguous text. The aim is natural division and also very roughly comparable text component sizes. In regard to the latter experimental design choice, very varied text component lengths are easily accommodated.

2. Then both text units and the word set are projected into a Euclidean space. Correspondence analysis allows us to do this. This projection method takes "profiles" of counts, or frequencies of occurrence, endowed with the χ^2 metric, into a Euclidean space. Both text units and words are projected into the same Euclidean space. All pairwise relationships – between text units, between words, and between both sets – are taken into account in this mapping of the χ^2 metric endowed space into the Euclidean metric endowed space.

3. Within each text, based on its Euclidean factor space representation, we then proceed to investigate how ultrametric it is. By design, the "semantic network" used and expressed by the Euclidean factor space is metric. How ultrametric it is is the question we raise.

4. In one study, we look at the words, and seek out ultrametrically related words.

In Section 9.2 we discuss how we quantify ultrametricity. In Section 9.3 we describe the semantic mapping methodology through correspondence analysis. This is the mapping of recorded or input data endowed with the χ^2 metric into a Euclidean, factor space. In this Euclidean space, we then pose the question: how ultrametric is the given space? In Section

9.4 we summarize and discuss experimental results. We characterize texts and collections of text, "fingerprinting" them in terms of inherent ultrametricity. In Section 9.5 we look within a text to determine just where ultrametricity arises.

9.2   Quantifying Ultrametricity

In the background paper [179] it was described how ultrametricity provides a representation (in this sense a model) of Matte Blanco's symmetric reasoning. Symmetric reasoning, as we have seen in the previous chapter, is associated with repressed or otherwise unconscious thought processes.

Before introducing our method of quantifying ultrametricity, we look at some other ways we could do so, albeit in a less satisfactory way (as we will argue).

9.2.1   Ultrametricity Coefficient of Lerman

The principle adopted in any constructive assessment of ultrametricity is to construct an ultrametric on data and see what discrepancy there is between input data and induced ultrametric data structure. Quantifying ultrametricity using a constructive approach is less than perfect as a solution, given the potential complications arising from known problems, for example, chaining in single link, and non-uniqueness, or even inversions, with other methods. The conclusion here is that the "measurement tool" used for quantifying ultrametricity itself occupies an overly prominent role relative to that which we seek to measure. For such reasons, we need an independent way to quantify ultrametricity.

Lerman's [137] H-classifiability index is as follows. From the isosceles triangle principle, given a distance d where $d(x,y) \neq d(y,z)$, we have $d(x,z) \leq \max\{d(x,y), d(y,z)\}$, and it follows that the largest and second largest of the numbers $d(x,y), d(y,z), d(x,z)$ are equal. Lerman's H-classifiability measure essentially looks at how close these two numbers (largest and second largest) are. So as to avoid the influence of the distribution of the distance values, Lerman's measure is based on ranks (of these distances) only. For further discussion of it, see [169].

There are two drawbacks with Lerman's index, although it is both insightful and inspiring. Firstly, ultrametricity is associated with $H = 0$ but non-ultrametricity is not bounded. In extensive experimentation, we found maximum values for H in the region of 0.24. The second problem with Lerman's index is that for floating-point coordinate values, especially in high dimensions, the strict equality necessitated for an equilateral triangle is nearly impossible to achieve. However, our belief is that approximate equilateral triangles are very likely to arise in important cases of high-dimensional spaces with data points at hypercube vertex locations. We would prefer therefore that the quantification of ultrametricity should "gracefully" take account of triplets which are "close to" equilateral. Note that for some authors, the equilateral case is considered to be "trivial" or a "trivial limit" [232]. For us, however, it is an important case, together with the other important case of ultrametricity (i.e. isosceles with small base).

9.2.2   Ultrametricity Coefficient of Rammal, Toulouse and Virasoro

The quantification of how ultrametric a data set is by Rammal et al. [206, 207] was influential for us in this work. The Rammal ultrametricity index is given by $\sum_{x,y}(d(x,y) - d_c(x,y))/\sum_{x,y} d(x,y)$ where d is the metric distance being assessed, and d_c is the subdominant ultrametric. The latter is also the ultrametric associated with the single link hierarchical clustering method. The Rammal et al. index is bounded by 0 (= ultrametric) and 1. As pointed out in Rammal et al. [206, 207], this index suffers from "the chaining effect and from sensitivity to fluctuations". The single link hierarchical clustering method, yielding the subdominant ultrametric, is, as is well known, subject to such difficulties.

9.2.3 Ultrametricity Coefficients of Treves and of Hartman

Treves [232] considers triplets of points giving rise to minimal, median and maximal distances. In the plot of $d_{\min}/d_{\max}$ against $d_{\mathrm{med}}/d_{\max}$, the triangle inequality, the ultrametric inequality and the "trivial limit" of equilateral triangles occupy definable regions.

Hartmann [93] considers $d_{\max} - d_{\mathrm{med}}$. Lerman [137] uses ranks in order to give (translation, scale, etc.) invariance to the sensitivity (i.e. instability, lack of robustness) of distances. Hartmann instead fixes the remaining distance $d_{\min}$.

We seek to avoid, as far as possible, lack of invariance due to use of distances. We seek to quantify both isosceles with small base configurations, as well as equilateral configurations. Finally, we seek a measure of ultrametricity bounded by 0 and 1.

9.2.4 Bayesian Network Modelling

Latent ultrametric distances were estimated by Schweinberger and Snijders [214] using a Bayesian and maximum likelihood approach in order to represent transitive structures among pairwise relationships. As they state: "The observed network is generated by hierarchically nested latent transitive structures, expressed by ultrametrics". Multiple, nested transitive structures are at issue. "Ultrametric structures imply transitive structures" and as an informal way to characterize ultrametric structures (arising from embedded clusters, comprising "friends" and "close friends"): "Friends are likely to agree, and unlikely to disagree; close friends are very likely to agree, and very unlikely to disagree."

However, issues in the statistical model-based approach to determining ultrametricity include that convergence to an optimal fit is not guaranteed and there can be an appreciable computational requirement. Our approach (to be described in the next subsection), in contrast, is fast and can be achieved through sampling which supposes that there is a homogeneous ultrametricity pertaining to the data used. If sampling is used (for computational reasons) then we assume that the text is "textured" in the same way throughout, or that it is sufficiently "unified". For one theme in regard to content, or one origin, or one author, such an assumption seems a reasonable one.

9.2.5 Our Ultrametricity Coefficient

We define a coefficient of ultrametricity termed α which is specified algorithmically as follows.

1. All triplets of points are considered, with a distance (by default, Euclidean) defined on these points. Since for a large number of points, n, the number of triplets, $n(n-1)(n-2)/6$, would be computationally prohibitive, we may wish to randomly (uniformly) sample coordinates ($i \sim \{1, \ldots, n\}, j \sim \{1, \ldots, n\}, k \sim \{1, \ldots, n\}$).

2. We check for possible alignments (implying degenerate triangles) and exclude such cases.

3. Next we select the smallest angle as less than or equal to 60 degrees. (We use the well-known definition of the cosine of the angle facing side of length x as

$(y^2+z^2-x^2)/2yz$.) This is our first necessary property for being a strictly isosceles (< 60 degrees) or equilateral ($= 60$ degrees) ultrametric triangle.

4. For the two other angles subtended at the triangle base, we seek an angular difference of strictly less than 2 degrees (0.03490656 radians). This condition is an approximation to the ultrametric configuration, based on an arbitrary choice of small angle. This condition is targeting a configuration that may not be exactly ultrametric but nonetheless is very close to ultrametric.

5. Among all triplets (1) satisfying our exact properties (2, 3) and close approximation property (4), we define our ultrametricity coefficient as the relative proportion of these triplets. Approximately ultrametric data will yield a value of 1. On the other hand, data that are non-ultrametric in the sense of not respecting conditions 3 and 4 will yield a low value, potentially reaching 0.

In summary, the α index is defined as follows. Consider a triplet of points that defines a triangle. If the smallest internal angle, a, in this triangle is less than or equal to 60 degrees, and, for the two other internal angles, b and c, if $|b - c| < 2$ degrees, then this triangle is an ultrametric one. We look for the overall proportion of such ultrametric triangles in our data.

The essential pseudo-code used is summarized in Annex 1. Of course software for this is available on the book's website.

9.2.6 What the Ultrametricity Coefficient Reveals

A wide range of case studies are used in [169] to explore this coefficient of ultrametricity. It is found that:

- the number of points (i.e. either words or text components), n, does not affect the value of the ultrametricity coefficient, α;

- ultrametricity as quantified in this way increases with sparsity of data encoding (e.g. word presences in text components);

- ultrametricity increases with dimensionality (of either word set or text component set);

- dimensionality and spatial (embedding space – each word in the text component space, and each text component in the word space) sparsity, combined, force the tendency towards ultrametricity, but the compounding of these two data properties is not as pronounced as one might have expected;

- and ultrametricity very noticeably increases with spatial dimensionality.

Furthermore, in [169] a connection is made with sparse forms of coding in regard to how complex stimuli are represented in the cortex. Among other implications, this points to the possibility that semantic pattern matching is best accomplished through ultrametric computation.

In regard to such ultrametric computation, search can benefit from prior ultrametric structuring – such as through inducing a hierarchical clustering on the data – and then nearest neighbour search can be shown to be achievable in constant worst-case computational time. This very powerful result is in keeping with the human ability to pattern-match in thought in what appears to be real time. As concluded in [169], it may be the case that human thinking is computationally efficient precisely because such computation is carried out in an ultrametric space.

So much for the background on the experimental work now to be presented.

With regard to Matte Blanco [33], the human thinking at issue is "unrepressed unconscious" thinking, expressing symmetrical reasoning, or more the symmetrical mode of being. This is one facet of the bi-logical system in the human mind process.

9.3 Semantic Mapping: Interrelationships to Euclidean, Factor Space

We employ correspondence analysis for metric embedding, followed by determination of the extent of ultrametricity, in factor space, based on the α coefficient of ultrametricity. Our motivation for using precisely this Euclidean embedding is as follows. Our input data are in the form of frequencies of occurrence. Now, a Euclidean distance defined on vectors with such values is not appropriate.

The χ^2 distance is an appropriate weighted Euclidean distance for use with such data [24, 171]. Consider texts i and i' crossed by words j. Let k_{ij} be the number of occurrences of word j in text i. Then, omitting a constant, the χ^2 distance between texts i and i' is given by $\sum_j (1/k_j)(k_{ij}/k_i - k_{i'j}/k_{i'})^2$. The weighting term is $1/k_j$. The weighted Euclidean distance is between the *profile* of text i, namely, k_{ij}/k_i for all j, and the analogous *profile* of text i'.

The following short subsections briefly summarize essential analytics framework aspects.

9.3.1 Correspondence Analysis: Mapping χ^2 into Euclidean Distances

As a dimensionality reduction technique correspondence analysis is particularly appropriate for handling frequency data. As an example of the latter, frequencies of word occurrence in text will be studied below.

The given contingency table (or numbers of occurrence) data is denoted $k_{IJ} = \{k_{IJ}(i,j) = k(i,j); i \in I, j \in J\}$. I is the set of text indexes, and J is the set of word indexes. We have $k(i) = \sum_{j \in J} k(i,j)$. We define $k(j)$ analogously, and $k = \sum_{i \in I, j \in J} k(i,j)$. Next, $f_{IJ} = \{f_{ij} = k(i,j)/k; i \in I, j \in J\} \subset \mathbb{R}_{I \times J}$. Similarly, f_I is defined as $\{f_i = k(i)/k; i \in I, j \in J\} \subset \mathbb{R}_I$, and f_J analogously. In this way numbers of occurrences are converted into relative frequencies.

The conditional distribution of f_J knowing $i \in I$, also termed the jth profile with coordinates indexed by the elements of I, is

$$f_J^i = \{f_j^i = f_{ij}/f_i = (k_{ij}/k)/(k_i/k); f_i \neq 0; j \in J\},$$

and likewise for f_I^j.

Note that the input data values here are non-negative reals; output factor projections (and contributions to the principal directions of inertia) will be reals. It may just be noted that the marginals (i.e. the I and J masses) are necessarily positive, and this can allow for imput data values to be negative (cf. [131, pp. 59–60]).

9.3.2 Input: Cloud of Points Endowed with the Chi-Squared Metric

The cloud of points consists of the couple: profile coordinate and mass. We have $N_J(I) = \{(f_J^i, f_i); i \in I\} \subset \mathbb{R}_J$, and again similarly for $N_I(J)$.

The moment of inertia is

$$M^2(N_J(I)) = M^2(N_I(J)) = \|f_{IJ} - f_I f_J\|^2_{f_I f_J} = \sum_{i \in I, j \in J} (f_{ij} - f_i f_j)^2 / f_i f_j. \qquad (9.1)$$

The term $\|f_{IJ} - f_I f_J\|^2_{f_I f_J}$ is the χ^2 metric between the probability distribution (empirically defined from the frequencies) f_{IJ} and the product of marginal distributions $f_I f_J$, with as centre of the metric the product $f_I f_J$. Decomposing the moment of inertia of the cloud $N_J(I)$ – or of $N_I(J)$ since both analyses are inherently related – furnishes the principal axes of inertia, defined from a singular value decomposition.

9.3.3 Output: Cloud of Points Endowed with the Euclidean Metric in Factor Space

From the initial frequencies data matrix, a set of probability data, f_{ij}, is defined by dividing each value by the grand total of all elements in the matrix. In correspondence analysis, each row (or column) point is considered to have an associated weight. The weight of the ith row point is given by $f_i = \sum_j f_{ij}$, and the weight of the jth column point is given by $f_j = \sum_i f_{ij}$. We consider the row points to have coordinates f_{ij}/f_i, thus allowing points of the same *profile* to be identical (i.e. superimposed). The following weighted Euclidean distance, the χ^2 distance, is then used between row points:

$$d^2(i,k) = \sum_j \frac{1}{f_j} \left(\frac{f_{ij}}{f_i} - \frac{f_{kj}}{f_k} \right)^2.$$

An analogous distance is used between column points.

The mean row point is given by the weighted average of all row points:

$$\sum_i f_i \frac{f_{ij}}{f_i} = f_j$$

for $j = 1, 2, \ldots, m$. Similarly the mean column profile has ith coordinate f_i.

We first consider the projections of the n profiles in $\mathbb{R}^m$ onto an axis, $\mathbf{u}$. This is given by

$$\sum_j \frac{f_{ij}}{f_i} \frac{1}{f_j} u_j$$

for all i (note the use of the scalar product here). For details on determining the new axis, $\mathbf{u}$, see [171].

The projections of points onto axis $\mathbf{u}$ were with respect to the $1/f_i$ weighted Euclidean metric. This makes interpreting projections very difficult from a human/visual point of view, and so it is more natural to present results in such a way that projections can be simply appreciated. Therefore *factors* are defined, such that the projections of row vectors onto factor ϕ associated with axis $\mathbf{u}$ are given by

$$\sum_j \frac{f_{ij}}{f_i} \phi_j$$

for all i. Taking

$$\phi_j = \frac{1}{f_j} u_j$$

ensures this and projections onto ϕ are with respect to the ordinary (unweighted) Euclidean distance.

An analogous set of relationships hold in $\mathbb{R}^n$ where the best-fitting axis, $\mathbf{v}$, is searched for. A simple mathematical relationship holds between $\mathbf{u}$ and $\mathbf{v}$, and between ϕ and ψ (the latter being the factor associated with axis or eigenvector $\mathbf{v}$):

$$\sqrt{\lambda}\psi_i = \sum_j \frac{f_{ij}}{f_i}\phi_j,$$

$$\sqrt{\lambda}\phi_j = \sum_i \frac{f_{ij}}{f_j}\psi_i.$$

These are termed *transition formulas*. Axes $\mathbf{u}$ and $\mathbf{v}$, and factors ϕ and ψ, are associated with eigenvalue λ, and best-fitting higher-dimensional subspaces are associated with decreasing values of λ (see [171] for further details).

In this work, ϕ_j are coordinates of words in the new, factor and Euclidean, space. The ψ_i are coordinates of text segments in the factor space. In the Euclidean, factor space, the transition formulas have the following interpretation. Each text point is the weighted average of its associated word points. Similarly, each word is located at the centre of gravity of its associated texts. In this way the factor space of the text segments and the factor space of the words furnish one semantic space.

9.3.4 Conclusions on Correspondence Analysis and Introduction to the Numerical Experiments to Follow

Some important points in relation to correspondence analysis for the analyses to follow are:

1. From numbers of occurrence data we always get (by design) a Euclidean embedding using correspondence analysis. The factors are embedded in a Euclidean metric.

2. Due to centring the data, the numbers of factors (i.e. number of non-zero eigenvalues) are given by one less than the minimum of the number of observations studied (indexed by set I) and the number of variables or attributes used (indexed by set J).

3. The number of dimensions in factor space may be less than full rank if there are linear dependencies present.

4. In the experiments to follow in the next section, we have $n < m$ always, implying that inherent (full-rank) dimensionality of the projected Euclidean factor space is $n - 1$.

5. We also take $m = 1000, 2000$ and the full attribute set (say, m_{tot}) in each case, where the attributes are ordered in terms of decreasing marginal frequency. In other words, we take the 1000 most frequent words to characterize our texts; then the 2000 most frequent words; and finally all words. Since $n < m$ it is not surprising that similar results are found irrespective of the value of m. The inherent, projected, Euclidean, factor space dimensionality is the same in each case (viz., $n - 1$).

6. From the previous remark (viz. that the results obtained for the $m = 1000, 2000$, and all most frequent words, are of the same inherent dimensionality) we motivate our use of these different characterizations of the text set by the need to study the stability of our results. We will show quite convincingly that our results are characteristic of the texts used, in each case, and are not "one-off" or arbitrary.

Some important points related to our numerical assessments below, in relation to data used, determination of ultrametricity coefficient, and software used, are as follows.

1. In line with one tradition of textual analysis associated with Benzécri's correspondence analysis (see [171, Chapter 5]) we take the unique full words and rank them in order of importance. Thus for the Brothers Grimm work, below, we find function words: "the", 19,696 occurrences; "and", 14,582 occurrences; "to", 7380 occurrences; "he", 5951 occurrences; "was", 4122 occurrences; and so on. The last three, with one occurrence each, are "yolk", "zeal", "zest".

2. The α ultrametricity coefficient is based on triangles. Now, with n graph nodes we have $O(n^3)$ possible triangles which is computationally prohibitive, so we instead sample. The means and standard deviations below are based on 2000 random triangle vertex realizations, repeated 20 times; hence, in each case, in total 40,000 random selections of triangles.

3. All text collections reported on in Section 9.4 are publicly accessible (and web addresses are cited). All texts were obtained by us in straight (ASCII) text format.

   The preparation of the input data was carried out with programs written in C, and available at www.correspondances.info (accompanying [171]). The correspondence analysis software was written in the R statistical software environment (www.r-project.org; again see [171]) and is available at this same web address. Some simple statistical calculations were carried out by us also in the R environment.

9.4 Determining Ultrametricity through Text Unit Interrelationships

We use in all over 900 short texts, given by short stories, or chapters, or short reports. All are in English. Unique words are determined through delimitation by white space and by punctuation characters with no distinction of upper and lower case. In all, over 1 million words are used in our studies of these texts.

We carried out some assessments of Porter stemming [204] as an alternative to use of whitespace- or punctuation-delimited words, without much difference in our findings.

9.4.1 Brothers Grimm

As a homogeneous collection of texts we take 209 fairy tales of the Brothers Grimm [196], containing 7443 unique (in total 280,629) space- or punctuation-delimited words. Story lengths were between 650 and 44,400 words.

To define a semantic context of increasing resolution we took the most frequent 1000 words, followed by the most frequent 2000 words, and finally all 7443 words. We constructed a cross-tabulation of numbers of occurrences of each word in each one of the 209 fairy tales. This led therefore to a set of frequency tables (contingency tables) of dimensions: 209×1000, 209×2000 and 209×7443. The factor space, of dimension $209 - 1 = 208$ (see Section 9.3.4), is Euclidean, so the correspondence analysis can be said to be a mapping from the χ^2 metric into a Euclidean metric space.

Table 9.1 (columns 4, 5) shows remarkable stability of the α ultrametricity coefficient results, and such stability will be seen in all further results to be presented below. In the table, means and standard deviations were calculated in each case from 2000 random

TABLE 9.1: Coefficient of ultrametricity, α. Input data: frequencies of occurrence matrices defined on the 209 texts crossed by 1000, 2000 and all 7443 words. The coefficient is based on factors; that is, factor projections resulting from correspondence analysis, with Euclidean distance used between each pair of texts in factor space, of dimensionality 208.

209 Brothers Grimm fairy tales				
Texts	Orig. Dim.	Factor Dim.	α, mean	α, std. dev.
209	1000	208	0.1236	0.0054
209	2000	208	0.1123	0.0065
209	7443	208	0.1147	0.0066

TABLE 9.2: Coefficient of ultrametricity, α. Input data: frequencies of occurrence matrices defined on the 266 texts crossed by 1000, 2000 and all 9723 words. The coefficient is based on factors; that is, factor projections resulting from correspondence analysis, with Euclidean distance used between each pair of texts in factor space. The dimensionality of the latter is necessarily $\leq 266 - 1$, adjusted for zero eigenvalues (= linear dependence).

266 Austen chapters or partial chapters				
Texts	Orig. Dim.	Factor Dim.	α, mean	α, std. dev.
266	1000	261	0.1455	0.0084
266	2000	262	0.1489	0.0083
266	9723	263	0.1404	0.0075

triangles, repeated 20 times (see Section 9.3.4). The ultrametricity is not high for the Grimm Brothers' data: we recall that an α value of 0 means no triangle is isosceles/equilateral. We see that there is very little ultrametric (hence hierarchical) structure in the Brothers Grimm data (based on our particular definition of ultrametricity/hierarchy).

9.4.2 Jane Austen

To further study stories of a general sort, we use some works of the English novelist, Jane Austen.

1. *Sense and Sensibility* [8], 50 chapters = files, chapter lengths from 1028 to 5632 words.

2. *Pride and Prejudice* [9], 61 chapters each containing between 683 and 5227 words.

3. *Persuasion* [10], 24 chapters, chapter lengths 1579 to 7007 words.

4. *Sense and Sensibility* split into 131 separate texts, each containing around 1000 words (i.e. each chapter was split into files containing 5000 or fewer characters). We did this to check on any influence by the size (total number of words) of the text unit used (and we found no such influence).

In all there were 266 texts containing a total of 9723 unique words. We looked at the 1000, 2000 most frequent, and all 9723 words to characterize the texts by frequency of occurrence.

Table 9.2, again displaying very stable α values, indicates that the Austen corpus is a little more ultrametric than the Grimm corpus (Table 9.1).

TABLE 9.3: Coefficient of ultrametricity, α. Input data: frequencies of occurrence matrices defined on the 50 texts crossed by 1000, 2000 and all 4261 words. The coefficient is based on factors; that is, factor projections resulting from correspondence analysis, with Euclidean distance used between each pair of texts in factor space. The dimensionality of the latter is necessarily $\leq 50 - 1$, with an additional adjustment made for one zero eigenvalue, implying linear dependence.

50 aviation accident reports				
Texts	Orig. Dim.	Factor Dim.	α, mean	α, std. dev.
50	1000	48	0.1338	0.0077
50	2000	48	0.1186	0.0058
50	4261	48	0.1154	0.0050

9.4.3 Air Accident Reports

We used air accident reports to explore documents with very particular technical vocabulary. The National Transport Safety Board (NTSB) aviation accident database [195] contains information about civil aviation accidents in the United States and elsewhere. We selected 50 reports. Examples of two such reports used by us: occurred Sunday, January 02, 2000 in Corning, AR, aircraft Piper PA-46-310P, injuries – 5 uninjured; occurred Sunday, January 02, 2000 in Telluride, TN, aircraft: Bellanca BL-17-30A, injuries – 1 fatal. In the 50 reports, there were 55,165 words. Report lengths ranged between approximately 2300 and 28,000 words. The number of unique words was 4261.

We give as an example the start of our 30th report: "On January 16, 2000, about 1630 eastern standard time (all times are eastern standard time, based on the 24 hour clock), a Beech P-35, N9740Y, registered to a private owner, and operated as a Title 14 CFR Part 91 personal flight, crashed into Clinch Mountain, about 6 miles north of Rogersville, Tennessee. Instrument meteorological conditions prevailed in the area, and no flight plan was filed. The aircraft incurred substantial damage, and the private-rated pilot, the sole occupant, received fatal injuries. The flight originated from Louisville, Kentucky, the same day about 1532."

In Table 9.3 we find ultrametricity values that are marginally greater than those found for the Brothers Grimm (Table 9.1). It could be argued that the latter, too, uses its own technical vocabulary. We would need to use more data to see if we can clearly distinguish between the (small) ultrametricity levels of these two corpora.

9.4.4 DreamBank

With dream reports (i.e. reports by individuals on their remembered dreams) we depart from a technical vocabulary, and instead raise the question as to whether dream reports can perhaps be considered as types of fairy tale or story, or even akin to accident reports. From the Dreambank repository [63, 69, 213] we selected the following collections:

(1) "Alta: a detailed dreamer", in period 1985–1997, 422 dream reports.
(2) "Chuck: a physical scientist", in period 1991–1993, 75 dream reports.
(3) "College women", in period 1946–1950, 681 dream reports.
(4) "Miami Home/Lab", in period 1963–1965, 445 dream reports.
(5) "The Natural Scientist", 1939, 234 dream reports.
(6) "UCSC women", 1996, 81 dream reports.

TABLE 9.4: Coefficient of ultrametricity, α. Input data: frequencies of occurrence matrices defined on the 384 texts crossed by 1000, 2000 and all 11,441, words. The coefficient is based on factors; that is, factor projections resulting from correspondence analysis, with Euclidean distance used between each pair of texts in factor space, of dimensionality $385 - 1 = 384$.

385 dream reports				
Texts	Orig.Dim.	FactorDim.	α, mean	α, sdev.
385	1000	384	0.1998	0.0088
385	2000	384	0.1876	0.0095
385	11441	384	0.1933	0.0087

To have reports of adequate length, we requested report sizes of between 500 and 1500 words. With this criterion, from (1) we obtained 118 reports, from (2) and (6) we obtained no reports, from (3) we obtained 15 reports, from (4) we obtained 73 reports, and finally from (5) we obtained 8 reports. In all, we used 214 dream reports, comprising 13696 words.

As an example, here is the start of our 100th report: "I'm delivering a car to a man – something he's just bought, a Lincoln Town Car, very nice. I park it and go down the street to find him – he turns out to be an old guy, he's buying the car for nostalgia – it turns out to be an old one, too, but very nicely restored, in excellent condition. I think he's black, tall, friendly, maybe wearing overalls. I show him the car and he drives off. I'm with another girl who drove another car and we start back for it but I look into a shop first – it's got outdoor gear in it – we're on a sort of mall, outdoors but the shops face on a courtyard of bricks. I've got something from the shop just outside the doors, a quilt or something, like I'm trying it on, when it's time to go on for sure so I leave it on the bench. We go further, there's a group now, and we're looking at this office facade for the Honda headquarters."

With the above we took another set of dream reports, from one individual, Barbara Sanders. A more reliable (according to [69]) set of reports comprised 139 reports, and a second comprised 32 reports. In all 171 reports were used from this person. Typical lengths were from about 2500 up to 5322 words. The total number of words in the Barbara Sanders set of dream reports was 107,791.

First we analysed all dream reports, furnishing Table 9.4. In order to look at a more homogeneous subset of dream reports, we then analysed separately the Barbara Sanders set of 171 reports, leading to Table 9.5. (Note that this analysis is on a subset of the previously analysed dream reports, Table 9.4). The Barbara Sanders subset of 171 reports contained 7044 unique words in all.

Compared to Table 9.4 based on the entire dream report collection, Table 9.5 shows, on average, higher ultrametricity levels. It is interesting to note that the dream reports, collectively, are higher in ultrametricity level than our previous values for α; and that the ultrametricity level is raised again when the data used relates to one person.

We carried out a preliminary study of James Joyce's *Ulysses*, comprising 304,414 words in total. We broke this text into 183 separate sequential files, comprising between approximately 1400 and 2000 words each. The number of unique words in these 183 files was found to be 28,649. The ultrametricity α values for this collection of 183 Joycean texts were found to be less than the Barbara Sanders values, but higher than the global set of all dream reports.

TABLE 9.5: Coefficient of ultrametricity, α. Input data: frequencies of occurrence matrices defined on the 171 texts crossed by 1000, 2000 and all 7044 words. The coefficient is based on factors; that is, factor projections resulting from correspondence analysis, with Euclidean distance used between each pair of texts in factor space, of dimensionality $171 - 1 = 170$.

171 Barbara Sanders dream reports				
Texts	Orig. Dim.	Factor Dim.	α, mean	α, std. dev.
171	1000	170	0.2250	0.0089
171	2000	170	0.2256	0.0112
171	7044	170	0.2603	0.0108

9.5   Ultrametric Properties of Words

9.5.1   Objectives and Choice of Data

The foregoing analyses have been based on text segments and their interrelationships. As noted earlier, however, correspondence analysis projects both text segments and words, both endowed initially with the χ^2 metric, into the one Euclidean space. As also observed, this Euclidean factor space takes all interrelationships into consideration. We stress too that we are *not* using a reduced-dimensionality approximation of the factor space, as is often done so as to filter out from the data what is considered to be noise. Instead we use the full Euclidean and factor space dimensionality because we wish to study the data as given to us but simply endowed with the usual (i.e. unweighted) Euclidean distance. (We also assume no recoding of the input data such as through complete disjunctive or fuzzy or other forms of coding which could turn the χ^2 distance right away into a Euclidean distance; see [171] for discussion of such input data recoding.)

In order to have a text that ought to contain vestiges of ultrametricity because of subconscious thinking, admittedly subconscious thinking that was afterwards reported on in a fully conscious way, we took the Barbara Sanders dream reports. In Section 9.4.4 we have seen how ultrametric we found this data to be. In a discussion of this data set Domhoff [62] notes that there is "astonishing consistency" shown in dreams such as these over long periods of time.

Taking a set of 139 of the Barbara Sanders dream reports, as used in Section 9.4.4, we used the 2000 most frequently occurring words in these dream reports including function words. Then we took 30 words to carry out some experimentation with their ultrametric properties. These are listed in Table 9.6. We selected these words to have some personal names, some words that could be metaphors for the commonplace or the fearful, and some words that could be commonplace and hence banal.

Two sets of experiments were carried out. For both experiments, the 30 selected words were given by their Euclidean space vectors resulting from the correspondence analysis, carried out on the 139 dream reports × 2000 words. So the 30 selected words are vectors in a space of dimensionality $\min(139 - 1, 2000 - 1) = 138$. In the first experiment the ultrametric triangles formed between triples solely on the 30-word set were determined. So for each word, the number of triangles checked was $1 \times (30 - 1) \times (30 - 2)/2 = 406$. In the second experiment, the ultrametric triangles formed between the selected word and all pairs of the full set of 2000 words were used. The number of triangles checked for each word was $1 \times (2000 - 1) \times (2000 - 2)/2 = 1,997,001$. However, some of these have overlapping points, implying zero distances. Rather than 1,997,001 triangles to be checked for each of the 2000 words, instead 1,996,997 involved no zero-valued distance.

9.5.2 General Discussion of Ultrametricity of Words

General discussion of Table 9.6 follows.

- Note the semantic similarity between "road" and "car", clearest when dealing with the 30-word set in isolation, rather than the 30-word set in the full 2000-word context.

- Similarly, note the semantic similarity between "balloon" and "balloons".

- Regarding the following words, our information is from [62]; and further discussion is in [64]).

- "Derek" ("H", high number of ultrametric relationships found with this word): the dreamer, Barbara Sanders, formerly had a relationship with him.

- "Mabel" ("L", relatively low number of ultrametric relationships): co-worker. The relatively low number of ultrametric relationships found was based on the full 2000-word set (135,192 cases); but when the restricted 30-word set alone was used in isolation a much larger relative number of 60 ultrametric cases was noted.

- "cat" ("H", high number of ultrametric relationships): Barbara Sanders has several cats, treats them well in real life, thinks of them as mistreated in dreams.

- "gun" ("H", high number of ultrametric relationships): Her dreams seem to imply that she used guns when young, but this was not in fact the case.

- "Howard" ("H", high number of ultrametric relationships): ex-husband.

- "horse" ("L", relatively low number of ultrametric relationships): she rides in dreams, fears in real life.

9.5.3 Conclusions on the Word Analysis

Derek, with whom Barbara Sanders formerly had a relationship, and Howard, an ex-husband, both figure relatively highly in terms of ultrametric relationships, as can be seen in Table 9.6. Admittedly these ultrametric-respecting triplets are few in number compared to the total number of these triplets (1,996,997 per word).

The distribution of the ultrametric-respecting triangles in a data set such as this allows us to assess the statistical significance of ultrametricity of any given word. Our approach is to determine the empirical distribution function (rather than, say, a stochastic graph model). The justification is to have a data-driven baselining rather than an *a priori* model for the data. Therefore we looked at the approximately 2 million triangles that are with reference to any word among the 2000 words retained.

Hence for this distribution we used approximately 4000 million triangles. With reference to the third column, therefore, of Table 9.6, the maximum number of ultrametric-respecting triangles with account taken of all 2000 words was found as 206,496. To determine this we checked all 2000 words. The minimum number of ultrametric-respecting triangles is 31,346. These correspond respectively to our α ultrametricity coefficients of 0.103403 and 0.015697.

Note that the results of Table 9.5 were based on the dream reports. While the word results are different, this just points to different ultrametricity properties in the two dual spaces. Our provisional conclusion in regard to the difference in ultrametricity properties

Selected words	# UM cases 300-word set (total triangles: 406)	# UM cases 2000-word set (total triangles: 1,996,997)	Previous col.: H(igh), L(ow) (defined by median)
Tyler	24	132,193	L
Jared	19	126,617	L
car	14	99,631	L
road	14	107,924	L
Derek	15	187,027	H
John	17	137,802	H
Jamie	24	130,304	L
Peter	48	134,052	L
arrow	21	133,917	L
dragon	24	170,157	H
football	18	127,036	L
Lance	22	166,112	H
room	5	65,332	L
bedroom	13	129,206	L
family	26	165,286	H
game	19	171,561	H
Mabel	60	135,192	L
crew	31	128,655	L
director	19	143,889	H
assistant	58	135,250	L
balloon	23	138,154	H
ship	18	154,960	H
balloons	23	147,757	H
pudgy	41	131,698	L
Valerie	17	161,231	H
dolly	20	140,355	H
cat	11	144,958	H
gun	20	166,147	H
Howard	28	172,760	H
horse	52	132,675	L

TABLE 9.6: Results found on the Barbara Sanders set of 139 dream reports for 30 selected words. "# UM cases" = number of ultrametric (triangle) cases. The numbers of ultrametric-respecting triangles were sought. Such triangles are either equilateral or isosceles with small base using Euclidean input data.

in the dual spaces is that it may be useful to experiment with content tagging (see the Hall/Van de Castle coding system [69]).

The measured ultrametricity of the word "Derek" (former relationship) is at the 73.887 percentile, implying a 26% chance of being bettered in this data. The measured ultrametricity of the word "Howard" (ex-husband) is at the 65.583 percentile.

Our objective in this word analysis has been to indicate the type of vantage points that can be opened up through the topology analysis that has been our focus in this work.

9.6 Concluding Comments on this Chapter

We studied a range of text corpora, comprising about 1000 texts or text segments, containing over 1.3 million words. We found very stable ultrametricity quantifications of the text collections, across numbers of most frequent words used to characterize the texts, and sampling of triplets of texts. Notable aspects of our data analysis include: full inherent dimensionality used; full set of words used too in many cases; and finally in Section 9.5, sampling was not used but rather exhaustive processing.

We found that in all cases (save, perhaps, the Brothers Grimm versus air accident reports) there was a clear distinction between the ultrametricity values of the text collections.

Some very intriguing ultrametricity characterizations were found in our work. For example, we found that the technical vocabulary of air accidents did not differ greatly in terms of inherent ultrametricity compared to the Brothers Grimm fairy tales. Secondly, we found that Jane Austen's novels were clearly distinguishable from the Grimm fairy tales. Thirdly, we found dream reports to have higher ultrametricity level than the other text collections.

Values of our α ultrametricity coefficient were small but revealing and valuable, in the sense of being consistent (i.e. with small variability) and discriminatory (i.e. between genres).

It is interesting to speculate on how one would exploit the "strands" or "threads" of ultrametricity, and hence hierarchical structure, that we find. We use these metaphors ("strands", "threads") with care because an ultrametric triangle possibly shares vertices with a non-ultrametric triangle.

All in all, however, we have presented excellent proof of concept that from empirical – textual – data we can determine measures of ultrametricity, or hierarchical symmetry. To that extent we have developed an operational procedure for ranking (at least as a good first stage of processing) manifestations of reasoning in terms of Matte Blanco's symmetric, on the one hand, and asymmetric, on the other hand, logic.

9.7 Annex 1: Pseudo-Code for Assessing Ultrametric-Respecting Triplet

Assumed: vectors i, j, k are in a Euclidean space.

- For all triplets i, j, k, consider their Euclidean distances, $d_1 = d(i, j)$, $d_2 = d(j, k)$, $d_3 = d(i, k)$.

- Set $\epsilon = 1.0e^{-10}$.

- Exclude near-zero distances: only if $(d_1 > \epsilon \ \& \ d_2 > \epsilon \ \& \ d_3 > \epsilon)$ do the following:

- Determine cosines of the three angles in the triangle using scalar product, denoted $\cdot$.

  $c_1 = (d_1 \cdot d_1 + d_2 \cdot d_2 - d_3 \cdot d_3)/(2.0 \cdot d_1 \cdot d_2)$

  $c_2 = (d_2 \cdot d_2 + d_3 \cdot d_3 - d_1 \cdot d_1)/(2.0 \cdot d_2 \cdot d_3)$

  $c_3 = (d_1 \cdot d_1 + d_3 \cdot d_3 - d_2 \cdot d_2)/(2.0 \cdot d_1 \cdot d_3)$

  Order these and we will take the case such that $c_1 \leq c_2 \leq c_3$.

- Wanting the largest cosine to correspond to an angle less than 60 degrees and greater than 0, implying that we have a sufficient condition for an isosceles with small base triangle, we require the following. Allowing less than or equal to 60 degrees encompasses the equilateral triangle case. Angle and cosine vary inversely.

- if $(c_3 \geq 0.5 \ \& \ c_3 < 1.0)$ then:

  Assess difference of angles. Note: 2 degrees = 0.03490656 radians.

  $a_1 = \arccos(c_1)$

  $a_2 = \arccos(c_2)$

  if ($|a_1 - a_2| < 0.03490656$) then we have we have an ultrametric-respecting triplet.

9.8 Annex 2: Bradley Ultrametricity Coefficient

Used in this book, a coefficient of ultrametricity is defined from the relative proportion of triplets of points that are approximately ultrametric (viz. there is equality to within 2 degrees of the base angles in an isosceles triangle with small base) or equilateral. For application of this work, see [170, 169, 172]. Some other coefficients of ultrametricity are discussed. In [36] another coefficient of inherent ultrametricity is defined. This is based on the following.

Given a finite metric space (X, d) with vertex set X, and distance $d(x, y)$, $x, y \in X$, the *Vietoris–Rips graph* (due to a 1927 publication by L. Vietoris; see [251]), Γ_ϵ, $\epsilon > 0$, is the set X with edge $(x, y) \in X \times X$ when $d(x, y) \leq \epsilon$. The following definition [36] is also an outcome from the foregoing discussion:

Distance d is an ultrametric if and only if, for all $\epsilon > 0$, the connected components of Γ_ϵ are complete subgraphs (cliques) of Γ_ϵ. The ultrametricity coefficient is then to be based on the measure μ for a graph Γ associated with (X, d), as follows: $0 \leq \mu(\Gamma) = \frac{b_0(\Gamma)}{c(\Gamma)} \leq 1$, where $b_0(\Gamma)$ is the number of connected components, and $c(\Gamma)$ is the number of complete subgraphs. Since a complete subgraph is a connected component, but not necessarily vice versa, it is clear that any complete subgraph is a subgraph of a connected component (the size of the latter is greater than or equal to the size of the former, the complete subgraph).

We have here consideration of the two extremes in a hierarchy of connected and complete components, that is, derived from, respectively, the single link and the complete link agglomerative hierarchical clustering criteria.

The topological ultrametricity index [36] is defined as $t(X, d) = \frac{1}{M} \int_0^M \mu(\Gamma_\epsilon) d\epsilon$, where M is the diameter of the space (X, d). We have that $t(X, d)$ is scale-invariant, and $t(X, d) \in (0, 1]$. If $d(X, d) = 1$ then d is an ultrametric. For all levels, ϵ, the number of connected components equals the number of complete subgraphs, then the single link hierarchical clustering is identical to the complete link hierarchical clustering. This identity of the hierarchies ensures that, for all ϵ, the components are complete and, therefore, connected. In the definition of $t(X,d)$ we require this identity.

While the definition of $t(X,d)$ above is a continuous function of ϵ, consider when we start with a hierarchical clustering, that is, a (potentially coerced) embedding in an ultrametric space. There are $n-1$ levels in the hierarchy, $n = |X|$, that is, agglomerations and therefore clusters or components are defined for distances $d_1, d_2, \ldots, d_{n-1}$, with $0 < d_1$. With piecewise continuity of μ between successive values of d_k, $1 \leq k \leq n-1$, and taking these d_k values to result from a complete link agglomerative hierarchical clustering, we have the expression

$$t(X,d) = \frac{1}{d_{n-1}} \sum_{k=1}^{k=n-1} \mu(\Gamma_{d_k})(d_k - d_{k-1}).$$

A few notes on implementation considerations follow. These notes are relevant for direct comparability and use of multiple hierarchical clusterings. We take Fisher's iris data [76] as an example, containing 150 observation vectors crossed by four features. Firstly, two of the iris flowers, 102 and 143, are identical in their feature value. Therefore distance $= 0$ implies aggregration of those row vectors. In general, distance $= 0$ permits singleton clusters only. Secondly, for $n = 150$ observation vectors, there are $n-1$ levels in the binary hierarchy. This is subject to unique agglomerative criteria values. While this is an assumption for observation vectors with real-value feature values, in the case of the iris data, where measured values are to one decimal place of precision (minimum value 0.1, maximum value 7.9), the unique levels are, for the complete link hierarchy, 124, and 110 for the single link hierarchy. (The implementation was `hclust` in R.) Rounding and other floating-point implementation issues are very relevant, although such issues can be accepted in practice since result and outcome reproducibility is guaranteed.

Implementation 1 of the *topological ultrametric index* [36], tUI1, is as follows.

1. tUI1 begins with the complete link and single link dendrograms. Thus we start with partially ordered sets that are, respectively, cliques (complete subgraphs) and connected components.

2. The levels of the complete link dendrogram (superscript c used) define the set $d_1^c, d_2^c, \ldots, d_{n-1}^c$. This set is now applied to both dendrograms. (Note that $d_{n-1}^s \leq d_{n-1}^c$, and usually $d_{n-1}^s \ll d_{n-1}^c$.)

3. The set of $n-1$ partitions are read off the complete link dendrogram; and the set of $n-1$ partitions are read off the single link dendrogram.

4. The sizes (i.e. number of clusters) of the partitions are determined. Thus, for d_k^c, for all k, this gives us $\mu(\Gamma_{d_k})$ = number of clusters in the single link (connected component) dendrogram, divided by the number of clusters in the complete link (clique or complete subgraph) dendrogram, at agglomeration criterion level d_k.

5. We next determine the difference in distances $d_{k+1} - d_k$, for $k = 0, 1, \ldots, n-1$. (Note that $d_0 = 0$ is associated with the trivial partition containing singleton clusters. Note also that d_{n-1} is associated with the trivial partition with one cluster of all observation vectors.)

6. tUI1 is the sum of elementwise product of the step 4 outcome and the step 5 outcome, divided by d_{n-1}^c.

Implementation 2 of the topological ultrametric index [36], tUI2, is as follows.

1. In step 2 of algorithm tUI1, we consider all d_k for the set k of all pairwise distances. Thus in this ordered set of distances, the maximum is $d_{n(n-1)/2}$. We take

Data set	Dimensions	tUI1	tUI2	tUI3
iris	150×4	0.619055	0.3752904	0.619055
Sanders	30×138	0.5323482	0.5360077	0.4076564
uniform	30×138	0.8514676	0.8409845	0.1093338

TABLE 9.7: Ultrametric index coefficients for three data sets, using three different implementations of the tUI ultrametricity index.

the ordered set of unique distances. (Were it the case of having non-redundant ultrametric distances on input, then there would be $n - 1$ such unique values, given the $n(n - 1)/2$ pairwise distances.)

2. Steps 3–6 take into account this set of distances.

Implementation 3 of the topological ultrametric index [36], tUI3, is based on the following [36]: $\mu(\Gamma)$ "can be viewed as a measure of how many ultrametric balls (maximal cliques) the clusters (connected components) are made up of." Thus here, we have the finding, for each class in the clink partition, of the indices of that class; then the finding of the class labels of those same indices in the slink partition. Then total homogeneity is tested, that is, identity of those class labels. If identity then the clink class is an slink class. Note that clink and slink singletons have such matching class labels. What we have here is that a clique (a clink class) is a constituent part of a connected component (an slink class).

1. Given the level, k, take the clink (complete link, clique) partition. For each class in the partition, determine if each member of that class is a member of the same class of the slink (single link) partition at that level. Could the number of these clink classes that are embedded in the one slink class. That is, therefore, each maximal clique in a connected component.

2. For the level, $\mu(\Gamma_k)$ is determined as the number of embedded cliques divided by the total number of connected components.

3. As above, $d_k - d_{k-1}$ is also determined.

4. Thus for each level there is the sum of μ times the difference in level distances.

5. The maximum distance, d_{n-1}, divides this sum, to yield the tUI3 coefficient.

6. Note that this implementation uses levels $k = 1, 2, \ldots, n - 1$, with level distance values $d_1^c, d_2^c, \ldots, d_{n-1}^c$.

Examples of outcomes are shown in Table 9.7. The Fisher iris data is directly used, with Euclidean distance. So too is the uniformly distributed data set. For the Sanders data, the following was done. As used in [181], these 139 dream reports were, firstly, characterized by presence/absence or frequency of occurrence of words used. A word corpus of 2000 words was derived from these dream reports (each comprising a few paragraphs of text). A correspondence analysis was carried out in order to take the dream report, and corpus, data clouds, endowed with the χ^2 metric (which provides for the principle of distributional equivalence of row and column profiles), into a factor space endowed with the Euclidean distance. From this space, 30 terms were selected. The majority of these terms were people's names. Therefore this data set comprised 30 terms crossed by their factor projection scores in 138 dimensions. (From input data of dimensions 139×2000, a correspondence analysis results in a factor space, for the dual, row and column, spaces, of dimensionsality $\min(139 - 1, 2000 - 1)$.)

10

Concluding Discussion on Software Environments

10.1 Introduction

For recent developments in the processing of vary large, Big Data, contexts, for distributed processing, and storage, the following from open source Apache, could be relevant: Spark, Hadoop, Hive. For all, there is considerable ongoing activity that allows direct or indirect interfacing with R. In many lines of inquiry and investigation, it may be beneficial to associate very different software platforms, with specific functionality. Therefore, given the systems at one's disposal, it might well be the case that IBM's Watson Analytics, or Tableau, are availed of for preliminary, exploratory visualization and display of data, and then that both information synthesis and data analysis are the main focus. For distributed processing, a key concept, maybe relating to NoSQL database storage, is as follows: key–value pairs, allowing the accumulation, in a distributed and scalable way, followed by the feeding into the component layers of the overall data processing carried out. This is the basis of the MapReduce programming or algorithmic model that is used in Apache Hadoop.

A system that is much favoured by social scientists and many others is SPAD (from Coheris Analytics). This system is based on visual programming, thereby supporting interactive analytics.

Whether in a complementary mode, or in a superior, alternative mode, key features of this book are therefore: rigour and mathematical underpinnings; consolidating and extending analytics frameworks relating to, for example, "computational thinking", "dialoguing with data", and "visualization and verbalization of data"; with wide coverage of application domains.

A key consideration is how hierarchy and symmetry are so fundamental to addressing and unravelling the complexity of systems, and for the practical and actionable understanding of the remarkable simplicity of very high-dimensional data, and, what is necessary to note, given any dual space relationship, what becomes the remarkable simplicity of massive data sets and streams.

It is also an expectation that future and further challenges and requirements can be addressed by hierarchy and symmetry, related to such areas as forensics, and potentially contexts where data rights management is relevant, and all such fields where the genealogy of information plays a role.

An interesting and useful commentary is made in [130, pp. 19, 20] that legacy software in use may well be a big restraint, "having the perverse effect of reifying thought at the same time as the procedures". Having referred there to SAS, the R package, FactoMineR, and then SPAD, the commentary ends with a quotation from Jean-Paul Benzécri in regard to inclusiveness of roles in analytics processes. Here a few small changes have been made to the quotation, to emphasize analytics over analysis, and how important all of this is for society and policy-making. "In analytics, numerous disciplines can and should collaborate. The role of mathematics, although essential, is modest in the sense that one can use, more or less exclusively, classical theorems or elementary demonstration techniques. But it must be the case that certain abstract conceptions enter into the spirits of the users, the specialists who collect the data, and who should orientate the analysis in accordance with the fundamental problems that are appropriate to their business, their policy-making, or their research and learning. ... We believe that it is reserved to data analytics to adequately express the laws of what are complex in essence (living beings, social bodies, ecosystems), and therefore that cannot be dissected without losing their very nature."

10.2   Complementary Use with Apache Solr (and Lucene)

The Classification Literature Automated Search Service was an annual bibliographic service made available to members of the Classification Society (formerly the Classification Society of North America) between 1971 and 2013. Data from Volumes 23–41, covering the period 1994–2013, are available on the book's website, for search and retrieval using Apache Solr, a search platform built on Apache Lucene for indexing. In this data set, there are 93,191 bibliographic records.

Solr can be, and was, used to select certain attributes for the bibliographic references, which then allowed the discipline, and time evolution, of content to be studied using correspondence analysis. A full Solr implementation is provided on the book's website.

Another set of data is available, also with a Solr implementation, to allow selection and extraction of relevant attributes or subsets of the data. This data set consists of 152,998 cooking recipes. For the correspondence analysis, a selection was made of 247 ingredients used by these recipes.

A few other areas are also discussed. These relate to Apache Lucene, for indexing; the open source Semantic Vectors, which can use random projections to expedite clustering, with k-means among the clustering algorithms supported; and Apache Mahout, in particular, for machine learning methods. The data used are from the publicly available 2014 UK Research Evaluation Framework relating to description of research impact. A total of 6637 submission documents are used. As a zip file, these are just under 20 MB in total file size.

Accompanying the software here, and these data sets, there is a 30-page report that describes various aspects of the processing and analytics that were carried out.

10.3   In Summary: Treating Massive Data Sets with Correspondence Analysis

10.3.1   Aggregating Similar or Identical Profiles Is Welcome

The principle of distributional equivalence has been noted in previous chapters. It means the following. Consider two elements of the attribute cloud, indexed by the set J, and with indices of these two elements, $j_1, j_2 \in J$. For given data x, we can define a function $x : I \times J \to \mathbb{R}+$ of the Cartesian product of the observation cloud, indexed by the set I, and the attribute cloud, indexed by the set J; this is thus a mapping of the cross-product of I and J, therefore of all elements i, j, with $i \in I$ and $j \in J$. The jth attribute profile is x_{ij}/x_j. Now, consider the profiles of attributes j and j' to be the same: $x_{ij}/x_j = x_{ij'}/x_{j'}$. Let these attributes be aggregated, that is, attributes j_1 and j_2 are replaced by j_0, so that $x_{ij_0} = x_{ij} + x_{ij'}$ and $x_{j_0} = x_j + x_{j'}$. From the distribution equivalence of the profiles we have not only equivalence of the j and j' profiles, but also their individual equivalence with the aggregated j_0 profile: $x_{ij_0}/x_{j_0} = x_{ij}/x_j = x_{i,j'}/x_{j'}$. This equality is approximated where there is near equivalence of the profiles.

That there is no effect when aggregating equivalent profiles is justified as follows. When we have a profile, j, and another identical profile, j', as points, these are co-located. From the χ^2 distance, any distance involving them will be identical to the distance with either one of them. If it is a matter of distances between attributes, then the zero distance of distributionally equivalent profiles is accounted for. Recall that the χ^2 distance between observations i and i' is $d^2(i, i') = \sum_{j \in J} \frac{1}{x_j} \left(x_{ij}/x_i - x_{i'j}/x_{i'} \right)^2$

Note, in regard to notation, that typically frequencies are worked on, that is, x_{ij}/x where $x = \sum_{i \in I} \sum_{j \in J} x_{ij}$. Clearly though, the foregoing can be expressed in terms of the given data, the matrix of terms ij.

In conclusion in this subsection, aggregation, when profiles are identical, is a very desirable processing action, potentially for interpretational reasons.

10.3.2 Resolution Level of the Analysis Carried Out

In Section 1.3, and in particular Section 1.3.1, the set of attributes used related to the upper level of a taxonomy of the attributes. Firstly, it is very clear that the number of such attributes leads to a lower-dimensional Euclidean distance endowed, factor space mapping. Furthermore, attributes that are at a finer resolution level (i.e. at a lower level in the taxonomy or concept hierarchy) can be retained as supplementary elements. These finer resolution level attributes can be projected into the analysis.

Carrying out analytics at such a resolution level means that the selected, lower resolution, level constitutes the main data table used. So our analysis is focused on that selected part of the data – let us assume, from the attribute set, J. We therefore have information focusing. Then, as noted, other parts of our data, for example, at finer resolution levels, can be projected into the factor space as supplementary elements.

We can express such information focusing as following. We define a small number of aggregates of either observations or attributes, and we carry out the analysis on them. Then we project the full set of observations and attributes into the factor space. A benefit of this, for interpretation, is that a low-dimensional factor space (i.e. using just the first, second, ... principal axes) is likely to be a very good approximation to the inertia of the clouds of observations, or (identically) to the cloud of attributes. As noted above, benefiting from such information focusing is especially relevant when our data set is taken as coming from, or being associated with, and ontology or concept hierarchy.

In Section 6.12, for interpretation reasons, aggregate tweets were examined. In other work, described in [184], about 12 million tweets were analysed, and for this, aggregation by day was used. This was carried out primarily in order to help with interpretation.

10.3.3 Random Projections in Order to Benefit from Data Piling in High Dimensions

Benefiting from data piling, or concentration, in very high-dimensional space was explored in detail in Chapter 5, starting especially in Section 5.2.5. Experimentally, it was observed how random projections can be both stable and robust. That is, in regard to seriating one's data, mapping one's data onto a unidimensional curve, or axis, that accounts for proximity relations in one's data.

We noted the benefits of uniformly distributed random axes. Such processing is motivated by the knowledge that the singleton clusters of a hierarchical clustering, that is, the leaf nodes of the hierarchical tree, can be perfectly scaled in one dimension. Otherwise expressed, a seriation of the set of observations is integrally or inherently associated with the hierarchical clustering.

Data piling is a known outcome of very high-dimensional data spaces (i.e. clouds of points).

10.3.4   Massive Observation Cardinality, Moderate Sized Dimensionality

In Chapter 6, the convergence of aggregated and consensus random projections was at issue. Association was made to the following clustering approaches. First, there is power iteration clustering. This can be considered in terms of the effect of eigendecomposition, or spectral decomposition, which is associated with diagonalization. A real, symmetric matrix, A is decomposed into eigenvectors and eigenvalues, such that $A = U\Lambda U^t$, where U is an orthogonal matrix, and Λ is a diagonal matrix containing the eigenvalues of A. Just subject to the matrix values being unique, a unique block diagonal structure can be formed by array permuting, that is, row and column permuting, which are the same permutation given that a real, symmetric matrix is at issue. Such a block diagonal matrix structure can be taken as constituting an early-stage approximation of the eigenvalue, and associated eigenvector, decomposition. Many applications of the association of block diagonal structure formation and spectral or eigendecomposition are to be found in [31, 193].

In Section 6.8, the correspondence analysis decomposition was noted (equation (6.2)): for frequencies f,

$$f_{ij} = f_i f_j \left(1 + \sum_{\alpha=1,\dots,n} \lambda_\alpha^{-\frac{1}{2}} F_\alpha(i) G_\alpha(j) \right).$$

This expression is analogous to what is termed, for principal components analysis, the Karhunen–Loève decomposition or transform. Other terms used for it include expansion and reconstitution formula.

Implementation of correspondence analysis can include finding the trivial first eigenvector and associated eigenvalue that corresponds to the first term on the right here. Usually it will be dropped from consideration. This so-called trivial eigenvector and eigenvalue has the following explanation: it defines the centre of gravity of a cloud of points, relative to the initially given data origin.

We wish, however, to consider making use of the trivial eigenvalue and eigenvector. That is, when this is warranted by the very great size of, say, the observation set or cloud of points indexed by I. Now, if our data are well approximated by the relation $f_{ij} = f_i f_j$, then it is to be noted that the right-hand terms are the marginal probability distributions. Next, suppose that we have data piling, due to massive cardinality. Therefore our point cloud becomes very concentrated. This data piling even approximates a single point. This is nice for us because it implies that the decomposition may approximate the product of an axis by a point.

An application of this is in [183], where chromosomes in plant research are at issue. Features of the DNA in each chromosome are up to around 600,000 in value. So we map these chromosome features onto the marginal distribution, and then use this as seriation for deriving (feature-related) local clusters.

To conclude this section, it may be stated that massive data sets, in moderately sized dimensionality, can be well approximated in the manner that has been described. From that, then, clustering properties are to be derived.

10.4 Concluding Notes

All that both motivates and justifies the data analytic methods that are relevant and perhaps constitute a *sine qua non* of contemporary data analytics is based on metric geometry and ultrametric topology. Geometry and topology both open up prospects for the analyst and support the line of reasoning that one will have in carrying out analyses. That is, whether the analysis has standard or established processing pipelines, or is interactive, proactive and enlightened by what one discovers. Geometry and topology underpin how we address the complexity of observed or measured, or otherwise derived, data, and their intricacies, assuming they are indeed non-trivial and challenging in a constructive, and perhaps even inspirational, way.

10.6 Concluding Notes

Bibliography

[1] D. Achlioptas. Database-friendly random projections. In *PODS '01: Proceedings of the Twentieth ACM SIGMOD-SIGACT-SIGART Symposium on Principles of Database Systems*, pages 274–281, New York, 2001. ACM.

[2] T.W. Adorno. Difficulties. In *Essays on Music*, Berkeley, 2002. University of California Press. Selected, with Introduction, Commentary, and Notes by R. Leppert, new translations by S.H. Gillespie.

[3] M.V. Altaisky and B.G. Sidharth. p-Adic physics below and above Planck scales. *Chaos, Solitons and Fractals*, 10(2–3):167–176, 1999.

[4] V. Anashin and A. Khrennikov. *Applied Algebraic Dynamics*. Walter de Gruyter, Berlin, 2009.

[5] E. Anderson. The irises of the Gaspé peninsula. *Bulletin of the American Iris Society*, 59.2–5, 1935.

[6] M.L. Anderson. Embodied cognition: a field guide. *Artificial Intelligence*, 149:91–130, 2003.

[7] Aristotle. *Poetics*. Penguin, 1997. Translated by M. Heath.

[8] J. Austen. *Sense and Sensibility*. 1811. http://www.pemberley.com/etext/SandS.

[9] J. Austen. *Pride and Prejudice*. 1813. http://www.pemberley.com/etext/PandP.

[10] J. Austen. *Persuasion*. 1817. http://www.pemberley.com/etext/Persuasion.

[11] V. Avetisov and A. Bikulov. Protein ultrametricity and spectral diffusion in deeply frozen proteins. *Biophysical Reviews and Letters*, 3:387–396, 2008.

[12] V.A. Avetisov, A.H. Bikulov, S.V. Kozyrev, and V.A. Osipov. p-Adic models of ultrametric diffusion constrained by hierarchical energy landscapes. *Journal of Physics A: Mathematical and General*, 35:177–189, 2002.

[13] P. Aylward. *Understanding Dublane and Other Masscares: Forensic Studies of Homicide, Paedophilia, and Anorexia*. Karnac, London, 2012.

[14] C. Bandt. Ordinal time series analysis. *Ecological Modelling*, 182:229–238, 2005.

[15] C. Bandt and B. Pompe. Permutation entropy: a natural complexity measure for time series. *Physical Review Letters*, 88:174102(4), 2002.

[16] C. Bandt and F. Shiha. Order patterns in time series. Technical report, Institute of Mathematics, Greifswald, 2005. Preprint 3/2005, www.math-inf.uni-greifswald.de/~bandt/pub.html.

[17] R.G. Baraniuk, V. Cevher, M.F. Duarte, and C. Hegde. Model-based compressive sensing, 2008. http://arxiv.org/abs/0808.3572.

[18] A. Barron, J. Rissanen, and B. Yu. The minimum description length principle in coding and modeling. *IEEE Transactions on Information Theory*, 44:2743–2760, 1998.

[19] M. Bécue-Bertaut, B. Kostov, A. Morin, and G. Naro. Rhetorical strategy in forensic speeches: Multidimensional statistics-based methodology. *Journal of Classification*, 31:85–106, 2014.

[20] R. Bellman. *Adaptive Control Processes: A Guided Tour*. Princeton University Press, Princeton NJ, 1961.

[21] J.J. Benedetto and R.L. Benedetto. A wavelet theory for local fields and related groups. *Journal of Geometric Analysis*, 14:423–456, 2004.

[22] R.L. Benedetto. Examples of wavelets for local fields. In C. Heil, P. Jorgensen, and D. Larson, editors, *Wavelets, Frames, and Operator Theory, Contemporary Mathematics Vol. 345*, pages 27–47. American Mathematical Society, Providence, RI, 2004.

[23] J. Benois-Pineau, A.Y. Khrennikov, and N.V. Kotovich. Segmentation of images in p-adic and Euclidean metrics. *Doklady Mathamtics*, 64:450–455, 2001.

[24] J.-P. Benzécri. *Analyse des Données, Volume 2, Correspondances*. Dunod, Paris, 2nd edition, 1976.

[25] J.-P. Benzécri. *Analyse des Données, Volume 1, Taxinomie*. Dunod, Paris, 2nd edition, 1979.

[26] J.-P. Benzécri. *Histoire et Préhistoire de l'Analyse des Données*. Dunod, Paris, 1982.

[27] J.-P. Benzécri. L'avenir de l'analyse des données (The future of data analysis). *Behaviormetrika*, 10(14):1–11, 1983.

[28] J.-P. Benzécri. *Correspondence Analysis Handbook*. Marcel Dekker, New York, 1992.

[29] J.P. Benzécri. L'approximation stochastique en analyse des correspondances. *Cahiers de l'Analyse des Données*, 7:387–394, 1982.

[30] J.P. Benzécri. Approximation stochastique, réseaux de neurones et analyse des données. *Cahiers de l'Analyse des Données*, 22:211–220, 1997.

[31] M.W. Berry, B. Hendrickson, and P. Raghavan. Sparse matrix reordering schemes for browsing hypertext. In *The Mathematics of Numerical Analysis, Lectures in Applied Mathematics Vol. 32*, pages 99–123. American Mathematical Society, Providence, RI, 1996.

[32] E. Bingham and H. Mannila. Random projection in dimensionality reduction: Applications to image and text data. In *KDD '01: Proceedings of the Seventh International Conference on Knowledge Discovery and Data Mining*, pages 245–250, New York, 2001. ACM.

[33] I. Matte Blanco. *The Unconscious as Infinite Sets: An Essay in Bi-Logic*. Karnac, London, 1998. Original version 1975. New foreword by Eric Rayner.

[34] J. Blasius and M. Greenacre, editors. *Visualization and Verbalization of Data*. Chapman & Hall/CRC Press, Boca Raton, FL, 2014.

[35] P. Bradley. Mumford dendrograms. *Computer Journal*, 53:393–404, 2010.

[36] P. Bradley. Finding ultrametricity in data using topology. *Journal of Classification*, 34, 2017.

[37] P.E. Bradley. On *p*-adic classification. *p-Adic Numbers, Ultrametric Analysis and Applications*, 1:271–285, 2009.

[38] L. Brekke and P.G.O. Freund. *p*-Adic numbers in physics. *Physics Reports*, 233:1–66, 1993.

[39] F.P. Brooks. Three great challenges for half-century-old computer science. *Journal of the ACM*, 50:25–26, 2003.

[40] J. Bruner. The narrative construction of reality. *Critical Inquiry*, 18:1–21, 1991.

[41] R. Buaba, A. Homaifar, M. Gebril, E. Kihn, and M. Zhizhin. Satellite image retrieval using low memory locality sensitive hashing in Euclidean space. *Earth Science Informatics*, 4:17–28, 2011.

[42] W.F. Buckley. Variations, review of Johann Sebastian Bach: Life and Work, by M. Geck, translated by J. Hargraves, Harcourt, 2006. *New York Times, Sunday Book Review*, 2006. 3 December.

[43] M. Burnett and J. Allison. *Everybody Comes to Rick's*. 1940. Screenplay, http://www.self.gutenberg.org/articles/everybody_comes_to_rick's.

[44] W. Chafe. *Discourse, Consciousness, and Time: The Flow and Displacement of Conscious Experience in Speaking and Writing*. University of Chicago Press, Chicago, 1994.

[45] W.L. Chafe. The flow of thought and the flow of language. In T. Givón, editor, *Syntax and Semantics: Discourse and Syntax, Volume 12*, pages 159–181. Academic Press, New York, 1979.

[46] P. Chakraborty. Looking through newly to the amazing irrationals. Technical report, 2005. arXiv: math.HO/0502049v1.

[47] J.-W. Chang and D.-S. Jin. A new cell-based clustering method for large, high-dimensional data in data mining applications. In *SAC '02: Proceedings of the 2002 ACM Symposium on Applied Computing*, pages 503–507, New York, 2002. ACM.

[48] H. Chernoff. The use of faces to represent points in *k*-dimensional space graphically. *Journal of the American Statistical Association*, 68:361–368, 1973.

[49] A. Clark. *Being There: Putting Brain, Body, and World Together Again*. MIT Press, Cambridge, MA, 1998.

[50] M. Clint and A. Jennings. The evaluation of eigenvalues and eigenvectors of real symmetric matrices by simultaneous iteration. *Computer Journal*, 13:76–80, 1970.

[51] P. Contreras and F. Murtagh. Fast, linear time hierarchical clustering using the Baire metric. *Journal of Classification*, 29:118–143, 2012.

[52] M. Costa, A.L. Goldberger, and C.-K. Peng. Multiscale entropy analysis of biological signals. *Physical Review E*, 71:021906(18), 2005.

[53] F. Critchley and W. Heiser. Hierarchical trees can be perfectly scaled in one dimension. *Journal of Classification*, 5:5–20, 1988.

[54] S. Dasgupta. Experiments with random projection. In *Proceedings of the 16th Conference on Uncertainty in Artificial Intelligence*, pages 143–151, San Francisco, 2000. Morgan Kaufmann.

[55] S. Dasgupta and A. Gupta. An elementary proof of a theorem of Johnson and Lindenstrauss. *Random Structures and Algorithms*, 22:60–65, 2002.

[56] B.A. Davey and H.A. Priestley. *Introduction to Lattices and Order*. Cambridge University Press, Cambridge, 2nd edition, 2002.

[57] S. Deegalla and H. Boström. Reducing high-dimensional data by principal component analysis vs. random projection for nearest neighbor classification. In *ICMLA '06: Proceedings of the 5th International Conference on Machine Learning and Applications*, pages 245–250, Los Alamitos, CA, 2006. IEEE Computer Society.

[58] F. Delon. Espaces ultramétriques. *Journal of Symbolic Logic*, 49:405–502, 1984.

[59] S.B. Deutsch and J.J. Martin. An ordering algorithm for analysis of data arrays. *Operations Research*, 19:1350–1362, 1971.

[60] A. Dijksterhuis and L.F. Nordgren. A theory of unconscious thought. *Perspectives on Psychological Science*, 1:95–109, 2006.

[61] A. Dix. The ultimate interface and the sums of life? *Interfaces*, 50:16, 2002.

[62] G.W. Domhoff. Using content analysis to study dreams: applications and implications for the humanities. In K. Bulkeley, editor, *Dreams: A Reader on the Religious, Cultural, and Psychological Dimensions of Dreaming*, pages 307–319. Palgrave, New York, 2002.

[63] G.W. Domhoff. *The Scientific Study of Dreams: Neural Networks, Cognitive Development and Content Analysis*. American Psychological Association, Washington, DC, 2003.

[64] G.W. Domhoff. Barb Sanders: Our best case study to date, and one that can be built upon, 2006. http://www2.ucsc.edu/dreams/Findings/barb_sanders.html.

[65] D.L. Donoho and J. Tanner. Neighborliness of randomly-projected simplices in high dimensions. *Proceedings of the National Academy of Sciences*, 102:9452–9457, 2005.

[66] B. Dragovich. *p*-Adic and adelic cosmology: *p*-adic origin of dark energy and dark matter. In A.Yu. Khrennikov, Z. Rakić, and I.V. Volovich, editors, *CP826, p-Adic Mathematical Physics: 2nd International Conference on p-Adic Mathematical Physics*, pages 25–42. American Institute of Physics, Melville, NY, 2006. https://arxiv.org/abs/hep-th/0602044.

[67] B. Dragovich and A. Dragovich. *p*-Adic modelling of the genome and the genetic code. *Computer Journal*, 53:432–42, 2010.

[68] B. Dragovich and L. Nešić. On space and time at very small distances. In I. Bikit, editor, *Proceedings of X Yugoslav Symposium on Nuclear and Elementary Particle Physics*, pages 102–105. 1996.

[69] DreamBank. Repository of dream reports, 2004. http://www.dreambank.net.

[70] D. Dutta, R. Guha, P.C. Jurs, and T. Chen. Scalable partitioning and exploration of chemical spaces using geometric hashing. *Journal of Chemical Information and Modeling*, 46:321–33, 2006.

[71] A. Edalat. Domains for computation in mathematics, physics and exact real arithmetic. *Bulletin of Symbolic Logic*, 3:401–452, 1997.

[72] A. Edalat. Domain theory and continuous data types, 2003. Lecture notes, http://www.doc.ic.ac.uk/~ae/teaching.html.

[73] J. Eliashberg, A. Elberse, and M.A.A.M. Leenders. The motion picture industry: critical issues in practice, current research, and new research directions. *Marketing Science*, 25:638–661, 2006.

[74] J. Eliashberg, S.K. Hui, and Z.J. Zhang. From storyline to box office: a new approach for green-lighting movie scripts. *Management Science*, 53:881–893, 2007.

[75] X. Z. Fern and C. Brodly. Random projection for high dimensional data clustering: A cluster ensemble approach. In *Proceedings of the Twentieth International Conference on Machine Learning*, Mcnlo Park, CA, 2003. AAAI Press.

[76] R. Fisher. The use of multiple measurements in taxonomic problems. *Annals of Eugenics*, 7:179–188, 1936.

[77] R. Foote. An algebraic approach to multiresolution analysis. *Transactions of the American Mathcmatical Society*, 357:5031–5050, 2005.

[78] R. Foote. Mathematics and complex systems. *Science*, 318:410–412, 2007.

[79] R. Foote, G. Mirchandani, D. Rockmore, D. Healy, and T. Olson. A wreath product group approach to signal and image processing: Part I – Multiresolution analysis. *IEEE Transactions on Signal Processing*, 48:102–132, 2000.

[80] R. Footc, G. Mirchandani, D. Rockmore, D. Healy, and T. Olson. A wreath product group approach to signal and image processing: Part II – Vonvolution, correlations and applications. *IEEE Transactions on Signal Processing*, 48:749–767, 2000.

[81] D. Fradkin and D. Madigan. Experiments with random projections for machine learning. In *KDD 2003: Proceedings of the Ninth ACM SIGKDD International Conference on Knowledge Discovery and Data Mining*, pages 517–522, New York, 2003. ACM.

[82] P. Frankl and H. Maehara. The Johnson-Lindenstrauss lemma and the sphericity of some graphs. *Journal of Combinatorial Theory, Series B*, 44(3):355–362, 1988.

[83] P.G.O. Freund. *p*-Adic strings and their applications. In B. Dragovich, A. Khrennikov, Z. Rakic, and I. Volovich, editors, *Proc. 2nd International Conference on p-Adic Mathematical Physics*, pages 65–73, Melville, NY, 2006. American Institute of Physics.

[84] L. Gajić. On ultrametric space. *Novi Sad Journal of Mathematics*, 31:69–71, 2001.

[85] G. Gan, C. Ma, and J. Wu. *Data Clustering Theory, Algorithms, and Applications*. Society for Industrial and Applied Mathematics, Philadelphia, 2007.

[86] B. Ganter and R. Wille. *Formal Concept Analysis: Mathematical Foundations*. Springer, Berlin, 1999. *Formale Begriffsanalyse: mathematische Grundlagen*, Springer, Berlin, 1996.

[87] A. Ganz. Let the audience add two plus two. They'll love you forever. The screenplay as a self teaching system. In J. Nelmes, editor, *Analysing the Screenplay*, pages 127–141. Routledge, New York, 2010.

[88] A. Ganz. Leaping broken narration: Ballads, oral storytelling and the cinema. In L. Khatib, editor, *Storytelling in World Cinema, Volume 1: Forms*. Wallflower Press, London, 2012.

[89] M.A. Giese and D.A. Leopold. Physiologically inspired neural model for the encoding of face spaces. *Neurocomputing*, 65/66:93–101, 2005.

[90] F.Q. Gouvêa. *p-Adic Numbers: An Introduction*. Springer, Berlin, 2003.

[91] P. Grabusts and A. Borisov. Using grid-clustering methods in data classification. In *PARELEC '02: Proceedings of the International Conference on Parallel Computing in Electrical Engineering*, pages 425–426, Los Alamitos, CA, 2002. IEEE Computer Society.

[92] P. Hall, J.S. Marron, and A. Neeman. Geometric representation of high dimensional, low sample size data. *Journal of the Royal Statistical Society B*, 67:427–444, 2005.

[93] A.K. Hartmann. Are ground states of 3d $\pm J$ spin glasses ultrametric? *Europhysics Letters*, 44:249–254, 1998.

[94] J. Hawkins. Why can't a computer be more like a brain? *IEEE Spectrum*, pages 17–22, April 2007.

[95] T.P. Hill. A statistical derivation of the significant-digit law. *Statistical Science*, 10(4):354–363, 1995.

[96] P. Hitzler and A. K. Seda. The fixed-point theorems of Priess-Crampe and Ribenboim in logic programming. *Fields Institute Communications*, 32:219–235, 2002.

[97] P. Indyk. Stable distributions, pseudorandom generators, embeddings and data stream computation. In *Foundations of Computer Science, FOCS 2000, Redondo Beach, CA*, pages 189–197, 2000.

[98] P. Indyk, A. Andoni, M. Datar, N. Immorlica, and V. Mirrokni. Locally-sensitive hashing using stable distributions. In *Nearest Neighbor Methods in Learning and Vision: Theory and Practice*, Cambridge, MA, 2006. MIT Press.

[99] A.K. Jain and R.C. Dubes. *Algorithms for Clustering Data*. Prentice Hall, Englewood Cliffs, NJ, 1988.

[100] A.K. Jain, M.N. Murty, and P.J. Flynn. Data clustering: a review. *ACM Computing Surveys*, 31:264–323, 1999.

[101] M.F. Janowitz. An order theoretic model for cluster analysis. *SIAM Journal on Applied Mathematics*, 34:55–72, 1978.

[102] M.F. Janowitz. Cluster analysis based on abstract posets. Technical report, Rutgers, The State University of New Jersey, 2005–2006. http://dimax.rutgers.edu/~melj.

[103] M. Jansen, G.P. Nason, and B.W. Silverman. Multiscale methods for data on graphs and irregular multidimensional situations. *Journal of the Royal Statistical Society B*, 71:97–126, 2009.

[104] S. C. Johnson. Hierarchical clustering schemes. *Psychometrika*, 32:241–254, 1967.

[105] W.B. Johnson and J. Lindenstrauss. Extensions of Lipschitz maps into a Hilbert space. In R. Beals, A. Beck, A. Bellow, and A. Hajian, editors, *Conference on Modern Analysis and Probability, Contemporary Mathematics Vol. 26*, pages 189–206. American Mathematical Society, Providence, RI, 1984.

[106] D.W. Jones. The Ternary Manifesto, 2012. Including "Standard ternary logic", "TerSCII: ternary standard code for information interchange", "Number representations for ternary computers", http://homepage.cs.uiowa.edu/~jones/ternary.

[107] S. Katok. *p-Adic Analysis Compared with Real*. American Mathematical Society, Providence, RI, 2007.

[108] N. Keiding and T.A. Louis. Perils and potentials of self-selected entry to epidemiological studies and surveys. *Journal of the Royal Statistical Society A*, 179:319–376, 2016.

[109] K. Keller and H. Lauffer. Symbolic analysis of high-dimensional time series. *International Journal of Bifurcation and Chaos*, 13:2657–2668, 2003.

[110] K. Keller, H. Lauffer, and M. Sinn. Ordinal analysis of EEG time series. *Chaos and Complexity Letters*, 2:247–258, 2007.

[111] K. Keller and M. Sinn. Ordinal analysis of time series. *Physica A*, 356:114–120, 2005.

[112] K. Keller and M. Sinn. Ordinal symbolic dynamics, 2005. Technical Report A-05-14, www.math.mu-luebeck.de/publikationen/pub2005.shtml.

[113] A. Khrennikov. *p-Adic Valued Distributions in Mathematical Physics*. Kluwer, Dordrecht, 1994.

[114] A. Khrennikov. *Non-Archimedean Analysis: Quantum Paradoxes, Dynamical Systems and Biological Models*. Kluwer, Dordrecht, 1997.

[115] A.Yu. Khrennikov. *Information Dynamics in Cognitive, Psychological, Social and Anomalous Phenomena*. Kluwer, Dordrecht, 2004.

[116] A.Yu. Khrennikov. Toward an adequate mathematical model of mental space: Conscious/unconscious dynamics on m-adic trees. *Biosystems*, 90:656–675, 2007.

[117] A.Yu. Khrennikov. Gene expression from polynomial dynamics in the 2-adic information space. *Proceedings of the Steklov Institute of Mathematics*, 265:131–139, 2009.

[118] A.Yu. Khrennikov. Modelling of psychological behavior on the basis of ultrametric mental space: Encoding of categories by balls. *p-Adic Numbers, Ultrametric Analysis, and Applications*, 2:1–20, 2010.

[119] A.Yu. Khrennikov. Dynamical processing of information in unconsciousness: ultrametric model. In *13th IEEE International Conference on Cognitive Informatics and Cognitive Computing*, Piscataway, NJ, 2014. IEEE.

[120] A.Yu. Khrennikov and S.V. Kozyrev. Pseudodifferential operators on ultrametric space and ultrametric wavelets. *Izvestia Mathematics*, 69:989–1003, 2005.

[121] A.Yu Khrennikov and S.V. Kozyrev. Wavelets on ultrametric spaces. *Applied and Computational Harmonic Analysis*, 19:61–76, 2005.

[122] S. King. *On Writing.* Pocket Books, New York, 2002.

[123] F. Klein. A comparative review of recent researches in geometry. *Bulletin of the New York Mathematical Society,* 2:215–249, 1893. Vergleichende Betrachtungen über neuere geometrische Forschungen, 1872, translated by M.W. Haskell.

[124] S. V. Kozyrev. Wavelet theory as p-adic spectral analysis. *Izvestiya: Mathematics,* 66:367–376, 2002.

[125] S. V. Kozyrev. Wavelets and spectral analysis of ultrametric pseudodifferential operators. *Sbornik: Mathematics,* 198:97–116, 2007.

[126] M. Krasner. Nombres semi-réels et espaces ultramétriques. *Comptes-Rendus de l'Académie des Sciences, Tome II,* 219:433–435, 1944.

[127] M. Kriegel and R. Aylett. Emergent narrative as a novel framework for massively collaborative authoring. In H. Prendinger, J. Lester, and M. Ishizuka, editors, *Intelligent Virtual Agents, Lecture Notes in Artificial Intelligence Vol. 5208,* pages 73–80, Berlin, 2008. Springer.

[128] V. Latora and M. Baranger. Kolmogorov-Sinai entropy rate versus physical entropy. *Physical Review Letters,* 82:520, 1999.

[129] R. Lauro-Grotto. The unconscious as an ultrametric set. *American Imago,* 64:535–543, 2007.

[130] B. Le Roux and F. Lebaron. Idées-clefs de l'analyse géométrique des données (Key ideas in the geometric analysis of data). In F. Lebaron and B. Le Roux, editors, *La Méthodologie de Pierre Bourdieu en Action: Espace Culturel, Espace Social et Analyse des Données,* pages 3–20. Dunod, Paris, 2015.

[131] B. Le Roux and H. Rouanet. *Geometric Data Analysis: From Correspondence Analysis to Structured Data Analysis.* Kluwer, Dordrecht, 2004.

[132] F. Lebaron. How Bourdieu "quantifed" Bourdieu: the geometric modelling of data. In K. Robson and C. Sanders, editors, *Quantifying Theory: Pierre Bourdieu.* Springer, Dordrecht, 2009.

[133] F. Lebaron and B. Le Roux, editors. *La Méthodologie de Pierre Bourdieu en Action: Espace Culturel, Espace Social et Analyse des Données.* Dunod, Paris, 2015.

[134] L. Lebart, A. Morineau, and K.M. Warwick. *Multivariate Descriptive Statistical Analysis.* Wiley, New York, 1984. Chapter 6, Direct Reading Algorithms.

[135] P. Legendre and L. Legendre. *Numerical Ecology.* Elsevier, Amsterdam, 3rd edition, 2012.

[136] D.A. Leopold, I.V. Bondar, and M.A. Giese. Norm-based face encoding by single neurons in the monkey inferotemporal cortex. *Nature,* 442:572–575, 2006.

[137] I.C. Lerman. *Classification et Analyse Ordinale des Données.* Dunod, Paris, 1981.

[138] L.A. Levin. The tale of one-way functions. *Problems of Information Transmission (Problemy Peredachi Informatsii),* 39:92–103, 2003. (Also arXiv.org/abs/cs.CR/0012023).

[139] A. Levy. *Basic Set Theory.* Dover, Mineola, NY, 2002. (Springer, 1979).

[140] M. Li and P. Vitányi. *An Introduction to Kolmogorov Complexity and Its Applications.* Springer, New York, 2nd edition, 1997.

[141] P. Li, T. Hastie, and K. Church. Very sparse random projections. In *KDD 2006: Proceedings of the 12th ACM SIGKDD International Conference on Knowledge Discovery and Data Mining*, volume 1, pages 287–296, New York, 2006. ACM.

[142] I. Liiv. Seriation and matrix reordering methods: An historical overview. *Statistical Analysis and Data Mining: The ASA Data Science Journal*, 3:70–91, 2010.

[143] F. Lin and W.W. Cohen. Power iteration clustering. In *Proc. 27th International Conference on Machine Learning, Haifa*, 2010.

[144] J. Lin and D. Gunopulos. Dimensionality reduction by random projection and latent semantic indexing. In *Proceedings of the Text Mining Workshop, 3rd SIAM International Conference on Data Mining*, Philadelphia, 2003. SIAM.

[145] A. Lohk, O. Tilk, and L. Võhandu. How to create order in large closed subsets of WordNet-type dictionaries. *Eesti Rakenduslingvistika Ühingu aastaraamat / Estonian Papers in Applied Linguistics*, 9:149–159, 2013.

[146] S. Lohr. Bill Gates, Andy Grove and Steve Jobs: The strategies they shared, 2015. *New York Times*, 12 May.

[147] S. Louchart, I. Swartjes, M. Kriegel, and R. Aylett. Purposeful authoring for emergent narrative. In *Interactive Storytelling: First Joint International Conference on Interactive Digital Storytelling, ICIDS 2008, Lecture Notes in Computer Science Vol. 5334*, pages 273–284, Berlin, 2008. Springer.

[148] S.C. Madeira and A.L. Oliveira. Biclustering algorithms for biological data analysis: a survey. *IEEE/ACM Transactions on Computational Biology and Bioinformatics*, 1:24–45, 2004.

[149] D.A. Madore. A first introduction to p-adic numbers. Revised 7 December 2000, http://www.madore.org/~david/math/padics.pdf.

[150] S.T. March. Techniques for structuring database records. *Computing Surveys*, 15:45–79, 1983.

[151] K. Marx. Capital Volume 1, Part I: Commodities and Money, Chapter 1: Commodities, Section 4, The fetishism of commodities and the secret thereof, 1867. Online edition, http://www.marxists.org/archive/marx/works/1867-c1/ch01.htm.

[152] W.T. McCormick, P.J. Schweitzer, and T.J. White. Problem decomposition and data reorganization by a clustering technique. *Operations Research*, 20:993–1009, 1982.

[153] R. McKee. *Story: Substance, Structure, Style, and the Principles of Screenwriting.* Methuen, York, 1999.

[154] R. Metzler, J. Klafter, and J. Jortner. Hierarchies and logarithmic oscillations in the temporal relaxation patterns of proteins and other complex systems. *Proceedings of the National Academy of Sciences*, 96:11085–11089, 1999.

[155] M.L. Miller, M. Acevedo Rodriguez, and I.J. Cox. Audio fingerprinting: Nearest neighbor search in high dimensional binary spaces. *Journal of VLSI Signal Processing Systems*, 41:285–291, 2005.

[156] B. Mirkin. *Mathematical Classification and Clustering*. Kluwer, Dordrecht, 1996.

[157] B. Mirkin. *Clustering for Data Mining*. Chapman & Hall/CRC, Boca Raton, FL, 2005.

[158] B. Mirkin and P. Fishburn. *Group Choice*. V.H. Winston, Washington, DC, 1979.

[159] M. Mitzenmacher. A brief history of generative models for power law and lognormal distributions. *Internet Mathematics*, 1:226–251, 2004.

[160] M.W. Moyer. The language of love: word usage predicts romantic attraction. *Scientific American*, 2011. 23 May.

[161] J. Murray. *Hamlet on the Hollodeck: The Future of Narrative in Cyberspace*. MIT Press, Cambridge, MA, 1998.

[162] F. Murtagh. Expected time complexity results for hierarchic clustering algorithms which use cluster centers. *Information Processing Letters*, 16:237–241, 1983.

[163] F. Murtagh. A survey of recent advances in hierarchical clustering algorithms. *Computer Journal*, 26:354–359, 1983.

[164] F. Murtagh. Complexities of hierarchic clustering algorithms: state of the art. *Computational Statistics Quarterly*, 1:101–113, 1984.

[165] F. Murtagh. Counting dendrograms: a survey. *Discrete Applied Mathematics*, 7:191–199, 1984.

[166] F. Murtagh. Structures of hierarchic clusterings: implications for information retrieval and for multivariate data analysis. *Information Processing and Management*, 20:611–617, 1984.

[167] F. Murtagh. *Multidimensional Clustering Algorithms*. Physica-Verlag, Heidelberg, 1985.

[168] F. Murtagh. Comments on: Parallel algorithms for hierarchical clustering and cluster validity. *IEEE Transactions on Pattern Analysis and Machine Intelligence*, 14:1056–1057, 1992.

[169] F. Murtagh. On ultrametricity, data coding, and computation. *Journal of Classification*, 21:167–184, 2004.

[170] F. Murtagh. Quantifying ultrametricity. In J. Antoch, editor, *COMPSTAT 2004 – Proceedings in Computational Statistics*, pages 1561–1568, Heidelberg, 2004. Physica Verlag.

[171] F. Murtagh. *Correspondence Analysis and Data Coding with Java and R*. Chapman & Hall/CRC, Boca Raton, FL, 2005.

[172] F. Murtagh. Identifying the ultrametricity of time series. *European Physical Journal B*, 43:573–579, 2005.

[173] F. Murtagh. The Haar wavelet transform of a dendrogram. *Journal of Classification*, 24:3–32, 2007.

[174] F. Murtagh. Editorial. *Computer Journal*, 51:612–614, 2008.

[175] F. Murtagh. The remarkable simplicity of very high dimensional data: application to model-based clustering. *Journal of Classification*, 26:249–277, 2009.

[176] F. Murtagh. Symmetry in data mining and analysis: a unifying view based on hierarchy. *Proceedings of Steklov Institute of Mathematics*, 265:177–198, 2009.

[177] F. Murtagh. The Correspondence Analysis platform for uncovering deep structure in data and information. *Computer Journal*, 53:304–315, 2010. 6th Annual Boole Lecture.

[178] F. Murtagh. On ultrametric algorithmic information. *Computer Journal*, 53:405–416, 2010.

[179] F. Murtagh. Ultrametric model of mind, I: Review. *p-Adic Numbers, Ultrametric Analysis and Applications*, 4:193–206, 2012.

[180] F. Murtagh. Ultrametric model of mind, II: Application to text content analysis. *p-Adic Numbers, Ultrametric Analysis, and Applications*, 4:207–221, 2012.

[181] F. Murtagh. Pattern recognition of subconscious underpinnings of cognition using ultrametric topological mapping of thinking and memory. *International Journal of Cognitive Informatics and Natural Intelligence*, 8:1 16, 2014.

[182] F. Murtagh. Ultrametric component analysis with application to analysis of text and of emotion, 2014. Preprint, http://arxiv.org/abs/1309.3611.

[183] F. Murtagh. Fast scaling followed by linear time clustering of massive data sets in moderate to low dimensions: exploiting the cloud dual spaces. 2016. Preprint, book's website.

[184] F. Murtagh. Semantic mapping: Towards contextual and trend analysis of behaviours and practices. In K. Balog, L. Cappellato, N. Ferro, and C. MacDonald, editors, *Working Notes of CLEF 2016 – Conference and Labs of the Evaluation Forum, Évora, Portugal, 5 8 September 2016*, pages 1207–1225, 2016. http://ceur-ws.org/Vol-1609.

[185] F. Murtagh and P. Contreras. Linear storage and potentially constant time hierarchical clustering using the Baire metric and random spanning paths. In A.F.X. Wilhelm and H.A. Kestler, editors, *Analysis of Large and Complex Data*, pages 43–52, Cham, 2016. Springer.

[186] F. Murtagh, G. Downs, and P. Contreras. Hierarchical clustering of massive, high dimensional data sets by exploiting ultrametric embedding. *SIAM Journal on Scientific Computing*, 30:707–730, 2008.

[187] F. Murtagh and A. Ganz. Pattern recognition in narrative: Tracking emotional expression in context. *Journal of Data Mining and Digital Humanities*, 2015. Online volume 2015, http://jdmdh.episciences.org/647.

[188] F. Murtagh, A. Ganz, and S. McKie. The structure of narrative: the case of film scripts. *Pattern Recognition*, 42:302–312, 2009.

[189] F. Murtagh, A. Ganz, S. McKie, J. Mothe, and K. Englmeier. Tag clouds for displaying semantics: The case of filmscripts. *Information Visualization Journal*, 9:253–262, 2010.

[190] F. Murtagh, A. Ganz, and J. Reddington. New methods of analysis of narrative and semantics in support of interactivity. *Entertainment Computing*, 2:115–121, 2011.

[191] F. Murtagh and A. Heck. *Multivariate Data Analysis*. Kluwer, Dordrecht, 1987.

[192] F. Murtagh, M. Pianosi, and R. Bull. Tracking and mapping Habermas's communicative action: A case study using Twitter social media. *Quality and Quantity*, 50:1675–1694, 2016.

[193] F. Murtagh, J.-L. Starck, and M. Berry. Overcoming the curse of dimensionality in clustering by means of the wavelet transform. *Computer Journal*, 43:107–120, 2000.

[194] Y. Neuman. *Introduction to Computational Cultural Psychology*. Cambridge University Press, Cambridge, 2014.

[195] NTSB. Aviation Accident Database and Synopses, National Transport Safety Board, 2003. http://www.landings.com.

[196] J.M. Ockerbloom. Grimm's Fairy Tales, 2003. http://www.cs.cmu.edu/~spok/grimmtmp.

[197] A.T. Ogielski and D.L. Stein. Dynamics on ultrametric spaces. *Physical Review Letters*, 55(15):1634–1637, 1985.

[198] K. O'Hara, R. Morris, N. Shadbolt, G.J. Hitch, W. Hall, and N. Beagrie. Memories for life: a review of the science and technology. *Journal of the Royal Society Interface*, 3:351–365, 2006.

[199] C. O'Neill. *Weapons of Math Destruction*. Crown/Archetype, Danvers, MA, 2016.

[200] A. Ostrowski. Über einige Lösungen der Funktionalgleichung $\phi(x) \cdot \phi(y) - \phi(xy)$. *Acta Mathematica*, 41:271–284, 1918.

[201] C.H. Papadimitriou. Mythematics: In praise of storytelling in the teaching of computer science and math. *Inroads – The SIGCSE Bulletin*, 35:7–9, 2003.

[202] N. H. Park and W. S. Lee. Statistical grid-based clustering over data streams. *SIGMOD Record*, 33:32–37, 2004.

[203] J.W. Pennebaker. *The Secret Life of Pronouns: What Our Words Say About Us*. Bloomsbury Press, London, 2012.

[204] M.F. Porter. An algorithm for suffix stripping. *Program*, 14:130–137, 1980.

[205] B. Quinn. Stephen Fry fans beg actor not to give up on Twitter. *The Observer*, 31 October 2009, http://www.theguardian.com/technology/2009/oct/31/stephen-fry-leave-twitter-fans.

[206] R. Rammal, J.C. Angles d'Auriac, and B. Doucot. On the degree of ultrametricity. *Journal de Physique – Lettres*, 46:945–952, 1985.

[207] R. Rammal, G. Toulouse, and M.A. Virasoro. Ultrametricity for physicists. *Reviews of Modern Physics*, 58:765–788, 1986.

[208] E. Rayner. *Unconscious Logic: An Introduction to Matte Blanco's Bi-Logic and Its Uses*. Routledge, London, 1995.

[209] J. Reddington, F. Murtagh, and D. Cowie. Computational properties of fiction writing and collaborative work. In A. Tucker and A.P.J.M. Siebes, editors, *Advances in Intelligent Data Analysis XII, Lecture Notes in Computer Science Vol. 8207*, pages 369–379, Heidelberg, 2013. Springer.

[210] H. Reiter and J.D. Stegeman. *Classical Harmonic Analysis and Locally Compact Groups*. Oxford University Press, Oxford, 2nd edition, 2000.

[211] J. Rissanen. The structure function and distinguishable models of data. *Computer Journal*, 49:657–664, 2006.

[212] W.H. Schikhof. *Ultrametric Calculus*. Cambridge University Press, Cambridge, 1984. (Chapters 18–21).

[213] A. Schneider and G.W. Domhoff. *The Quantitative Study of Dreams*. 2004. http://dreamresearch.net.

[214] M. Schweinberger and T.A.B. Snijders. Setting in social networks: A measurement model. *Sociological Methodology*, 33:307–342, 2003.

[215] A.K. Seda and P. Hitzler. Generalized distance functions in the theory of computation. *Computer Journal*, 53:443–464, 2010.

[216] J. Séguéla and G. Saporta. A comparison between latent semantic analysis and correspondence analysis, 2011. Presentation, CARME Conference, http//carme2011.agrocampus-ouest.fr/slides/Seguela_Saporta.pdf.

[217] R. Sibson. SLINK: an optimally efficient algorithm for the single link cluster method. *Computer Journal*, 16:30–34, 1973.

[218] J.H. Silverman. p-Adic numbers. 6 April 2010, 17 pages, http://www.math.brown.edu/~jhs/MA0156/NTUpadic.pdf.

[219] H.A. Simon. *The Sciences of the Artificial*. MIT Press, Cambridge, MA, 1996.

[220] N.J.A. Sloane. OEIS – On-Line Encyclopedia of Integer Sequences. Technical report, 2006. http://www.oeis.org (number sequence referenced by A000111).

[221] A.P. Stakhov. Brousentsov's ternary principle, Bergman's number system and ternary mirror-symmetrical arithmetic. *Computer Journal*, 45:221–236, 2002.

[222] A.P. Stakhov. *The Mathematics of Harmony. From Euclid to Contemporary Mathematics and Computer Science*. World Scientific, Singapore, 2009.

[223] J.-L. Starck, F. Murtagh, and A. Bijaoui. *Image Processing and Data Analysis: The Multiscale Approach*. Cambridge University Press, Cambridge, 1998.

[224] J.-L. Starck, F. Murtagh, and J.M. Fadili. *Sparse Image and Signal Processing: Wavelets and Related Geometric Multiscale Analysis*. Cambridge University Press, New York, 2nd edition, 2015.

[225] D. Steinley. k-means clustering: a half-century synthesis. *British Journal of Mathematical and Statistical Psychology*, 59:1–34, 2006.

[226] D. Steinley and M.J. Brusco. Initializing K-means batch clustering: a critical evaluation of several techniques. *Journal of Classification*, 24:99–121, 2007.

[227] I. Swartjes and M. Theune. Iterative authoring using story generation feedback: debugging or co-creation? In *Interactive Storytelling: Second Joint International Conference on Interactive Digital Storytelling*, pages 62–73, Berlin, 2009. Springer. Lecture Notes in Computer Science Vol. 5915.

[228] I. Swartjes and M. Theune. Late commitment: virtual story characters that can frame their world, 2009. Internal report, University of Twente, Centre for Telematics and Information Technology.

[229] A. Szalay, J. Gray, G. Fekete, P. Kunszt, P. Kukol, and A. Thakar. Indexing the sphere with the hierarchical triangular mesh, 2005. Microsoft Research Technical Report MSR-TR-2005-123.

[230] Y. Terada. Clustering for high-dimension, low-sample size data using distance vectors, 2013. http://arxiv.org/abs/1312.3386.

[231] M. Theune, S. Faas, A. Nijholt, and D. Heylen. The virtual storyteller. *SIGGROUP Bulletin*, 23:20–21, 2002.

[232] A. Treves. On the perceptual structure of face space. *BioSystems*, 40:189–196, 1997.

[233] The Tripatorium. The Faces of Sydney, Australia, 2010. http://thetripatorium.com/pictures/detail/the_faces_of_sydney_austrailia.

[234] W-K Tung. *Group Theory in Physics*. World Scientific, Philadelphia, 1985.

[235] I. van Mechelen, H.-H. Bock, and P. De Boeck. Two-mode clustering methods: a structured overview. *Statistical Methods in Medical Research*, 13:363–394, 2004.

[236] C.J. van Rijsbergen. *The Geometry of Information Retrieval*. Cambridge University Press, Cambridge, 2004.

[237] A.C.M. van Rooij. *Non-Archimedean Functional Analysis*. Marcel Dekker, New York, 1978.

[238] S.S. Vempala. *The Random Projection Method*. American Mathematical Society, Providence, RI, 2004.

[239] I.V. Volovich. p-Adic string. *Classical Quantum Gravity*, 4:L83–L87, 1987.

[240] I.V. Volovich. Number theory as the ultimate physical theory. *p-Adic Numbers, Ultrametric Analysis and Applications*, 2(1):77–87, 2010. Preprint No. TH 4781/87, CERN, Geneva, 1987.

[241] C.S. Wallace and D.L. Dowe. Minimum description length and Kolmogorov complexity. *Computer Journal*, 42:270–293, 1999.

[242] W. Weckesser. Symbolic dynamics in mathematics, physics, and engineering, based on a talk by N. Tuffilaro. Technical report, 1997. http://www.ima.umn.edu/~weck/nbt/nbt.ps.

[243] H. Weyl. *Symmetry*. Princeton University Press, Princeton, NJ, 1983.

[244] E. White. Aristotle, drama and the craft of reality TV. 22 March 2001, http://www.medialifemagazine.com/news2001/index.html.

[245] H.P. Wolf. Chernoff faces and spline interpolation, 2004. http://www.wiwi.uni-bielefeld.de/~wolf/software/R-wtools/faces/faces.pdf.

[246] P. Wu, S.C.H. Hoi, N. D. Dung, and H. Ying. Randomly projected KD-trees with distance metric learning for image retrieval. In *International Conference on Multimedia Modeling (MMM2011)*. Taipei, Taiwan, 2011.

[247] R. Xu and D. Wunsch. Survey of clustering algorithms. *IEEE Transactions on Neural Networks*, 16:645–678, 2005.

[248] R. Xu and D.C. Wunsch. *Clustering*. IEEE Press, Piscataway, NJ, 2008.

[249] W. Yan, U. Brahmakshatriya, Y. Xue, M. Gilder, and B. Wise. p-PIC: parallel power iteration clustering for big data. *Journal of Parallel and Distributed Computing*, 73:352–359, 2013.

[250] M.P. Young and S. Yamane. Sparse population coding of faces in the inferotemporal cortex. *Science*, 256:1327–1331, 1992.

[251] A. Zomorodian. Fast construction of the Vietoris-Rips complex. *Computers and Graphics*, 34:263–271, 2010.

Index